B

D1632054

An Introduction to Geology

An Introduction to Geology

BRIAN LEE

The Crowood Press

First published in 1987 by
The Crowood Press
Ramsbury, Marlborough
Wiltshire SN8 2HE

British Library Cataloguing in Publication Data

Lee, Brian
    An Introduction to Geology
    1. Geology
    I. Title
    551   QE26.2
    ISBN 0 946284 53 9

Acknowledgements

The photographs on page 116 and page 117 (top), and information
about their lead-zinc mine at Navan, Ireland, were generously
supplied by Tara Mines Ltd. British Petroleum kindly provided the
photograph of their Wytch Farm oil prospect on page 123. Permission
was granted by Tarmac Roadstones (North West) Ltd to take photographs in their Helwith Bridge quarries. Mr G. Hanley of Stump
Cross Caverns, North Yorkshire, was similarly helpful in permitting
photography of the caverns. Dr Ron Freethy read the manuscript and
his wife Marlene was, as ever, an efficient and imperturbable typist.

Typeset by Alacrity Phototypesetters,
Banwell Castle, Weston-super-Mare
Printed in Great Britain

Contents

Introduction

Just a few hundred years ago man's knowledge of the earth's structure and history was based upon a mixture of erroneous supposition and naïve superstition. For instance, the Cornish tin miners believed that the ore which they had extracted would, in time, be replaced by a process of spontaneous regeneration. In the Middle Ages, natives of the Whitby district of Yorkshire did a roaring trade in fake religious relics by collecting coiled fossil ammonites, which the pious believed to be the remains of snakes petrified by St Hilda, the local saint. To enhance the illusion, skilled carvers shaped the end of the outer coil into a serpent's head.

Nowadays our knowledge is a little more advanced, but although we have exploited the earth's resources for thousands of years, geology is still a relatively new science. There are many unexplained phenomena to be investigated, and large areas of the earth's surface have yet to be thoroughly surveyed. Our deepest boreholes merely scratch the planet's surface and, although we have developed considerable expertise at finding oil and metallic minerals, we are unable to agree about their formative processes.

Geology is the key to our future prosperity. It is the provider of raw materials for the voracious appetite of the consumer society. Civil engineering projects cannot proceed until geologists have assessed the reliability of underlying structures. Geology has even played its part in war. During World War II geologists were sent to survey the proposed sites of the Normandy landings in order to check that the terrain was suitable for heavy, tracked vehicles. Geology helps the naturalist, countryman or town dweller to appreciate the nature of his surroundings and to realise how the landscape has developed and why it supports a particular ecological system. It will also play a major role in our exploration of the solar system. Apart from all this, however, the subject has an addictive fascination. Anyone who has seen the play of light in a crystal of labradorite, found a gleaming, silvery cube of galena in a mineral vein, or chanced upon a perfectly preserved fossil, will be 'hooked' for life.

This book provides a sequence of up-to-date geological information for naturalists, and is a comprehensive introduction to the subject for beginners and students of geology and environmental studies. Beginning with a chapter on minerals – the basic raw materials of our planet – it goes on to describe how life developed, the various types of rock, scenery (which is dependent on the nature and structure of the underlying rocks), and man's use and extraction of raw materials and the environmental implications of this.

1 Minerals

To most people the names of a few gem minerals are far more familiar than those of the common minerals of the earth's crust. Yet diamonds, emeralds, sapphires and rubies are only found in great abundance in a jeweller's shop window, whilst the common minerals – quartz, micas, feldspars, olivines, pyroxenes, amphiboles, calcite, dolomite, rock salt, gypsum, garnet, apatite, magnetite and pyrite – are all widely distributed. Unfortunately, most of the important economic minerals are absent from this list as they usually occur with restricted distribution in localised mineral fields which cover only a small area of the earth's surface. Although there are many artificially imposed divisions in the classification of minerals, it is best to regard them as the building bricks of the earth's crust or, in the case of vein materials, as the mortar between the bricks.

Before the earth began to cool it was a revolving sphere of hot magma in space. As cooling took place, the heavier metallic minerals migrated towards the centre and the lighter compounds and elements floated to the surface as a hot magmatic scum – an ocean of fiery lava. Eventually, as the surface began to cool, the minerals with the higher melting points began to crystallise from the hot mixture. The first discrete minerals crystallised from hot magma several billion years ago. In Britain, some of these earliest rocks have been identified in the Northern Highlands of Scotland.

The structure of the rocks was determined by the chemical composition of the magma and by its rate of cooling. Slow cooling produced a coarsely crystalline structure where the individual mineral crystals may be seen without optical instruments. Quick cooling resulted in a fine-grained or glassy-textured rock with microscopically small crystals.

As the earth's crust matured, thickened, cooled and cracked, the seeds of change were sown. Deep below the solidified surface skin, which was beginning to wrinkle like the skin of a shrivelling apple, volcanic intrusions of magma occurred. Molten lava, gases and vapours were forced upwards through cracks and fissures. Hot mineralised solutions were injected into the older rocks, where they cooled and crystallised to produce the beautiful shapes which are treasured by mineral collectors. Some solutions actually forced their way into the porous structures of older rocks, cooling and solidifying within them. This caused chemical reactions to take place, and the structure and chemical composition of the rocks were drastically changed. This type of mineral deposit is known as a metasomatic replacement deposit. Other changes were caused by hot gases produced from consolidating magmatic intrusions. Such minerals were classified as pneumatolytic deposits. The processes of rock and mineral formation are still continuing many miles below the surface and sometimes on the surface, due to volcanoes or hot geysers.

Granite is one of the most abundant igneous rocks of the crust. Analysis of a typical granite would reveal that it contains large amounts of quartz (possibly 70 to 80 per cent), with smaller quantities of feldspar and mica plus traces of other minerals. However, not all rocks of this composition could be classified as granite. Other factors, including mode of formation and crystal structure, are important in identification. A good working knowledge of minerals is one of the prerequisites to becoming a competent geologist.

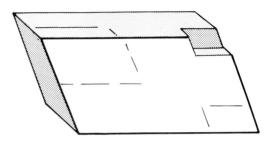

Identification

Minerals are usually identified by comparing their physical properties. The most useful properties for purposes of comparison are: hardness, cleavage, fracture, shape, colour, lustre, and specific gravity.

Hardness

Although there are various complex tests requiring sophisticated equipment, one of the simplest means of determining the approximate hardness of a mineral sample is to compare its hardness with minerals of known hardness from Moh's Scale: talc – 1, gypsum – 2, calcite – 3, fluorite – 4, apatite – 5, microcline (feldspar) – 6, quartz – 7, topaz – 8, corundum – 9, diamond – 10.

If these type minerals are not available, it is worth remembering that your fingernail has a hardness of 2.5 and a copper coin has a hardness of 3.5. Knife blades usually register a hardness of 5.5, with window glass slightly harder. Good quality steel is harder still and may be scratched by quartz only with difficulty, yet progressively more easily by topaz, corundum and diamond.

Cleavage

Cleavage is defined as the way in which a mineral disintegrates when subjected to pressure or a blow. Some minerals exhibit

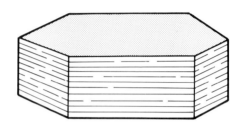

Top: calcite cleaves into smaller rhombs.
Bottom: a mica crystal cleaves into flakes.

no definite cleavage patterns, whereas others cleave into miniature replicas of their original form. Galena, for instance, remains cubic and tabular crystals ('books') of mica cleave into even thinner fragments. The cleavage of mica is uni-directional, while the cleavage of galena into small cubes is three-directional. The quality of cleavage shown by a mineral specimen is described as perfect, good, fair or poor.

Fracture

Fracture is seen when a mineral will not split cleanly on any cleavage plane, and so irregular breakage results. Quartz and glass are good examples of conchoidal fracture. Look closely at the edge of a broken window pane. It should be possible to see signs of shell-like, rounded fracture patterns. Asbestos, which is a greenish, woolly, serpentine mineral, fractures into fibrous threads.

Shape

The shape or crystal habit of a mineral may be distorted by the conditions in which it was formed, and is also influenced by the crystal system (see below). Shape may be related to cleavage and fracture. In some cases, such as in some replacement deposits, crystals may be completely malformed or even take the shape of another mineral. For example, where hematite has

replaced calcite its crystals will conform to the habit of calcite rather than hematite shape.

There are three basic habits: granular, prismatic and lamellar. Many minerals are found as granular, sugary masses with lots of ill-assorted grains or badly formed crystals cemented together. Prismatic minerals are those which form long, columnar, needle-like or fibrous crystals. Quartz and asbestos are at either end of the prismatic scale. Lamellar minerals are found in a 'platey' form.

Other minerals form aggregates or large masses, often appearing to radiate from a central area. Aggregates of prismatic mine-

Hexagonal quartz crystals.

rals form the most beautiful masses of crystals. Amethyst and rose quartz aggregates make particularly good exhibition pieces and command high prices. Earthy minerals may form botryoidal masses of gland-like shape. Kidney iron ore, hematite, is very aptly named. Manganese minerals tend to form dendritic patterns – tree-branch shapes which seem to be painted on to the rock surfaces.

Colour

Colour is a variable property in many minerals. There are notable exceptions, for example the ideochromatic minerals, which are invariably the same colour. The two carbonate ores of copper, malachite (green) and azurite (blue), are splendid examples of ideochromacy and are always the same colour.

The majority of minerals, however, may occur in a range of colours. These are known as allochromatic minerals. Quartz is a typical allochromatic mineral, and there are few colour varieties in which it does not occur. Fluorspar also exhibits many colour varieties.

It is not generally realised that sapphire and ruby are different colour varieties of the same mineral. Both are forms of corundum or aluminium oxide.

It is thought that the coloration of several minerals which may commonly be found as white or colourless, is caused by minute amounts of rare earth elements such as lanthanum. In other cases, colourless minerals are tinted by staining or impurities from nearby ideochromatic minerals. At the beginning of the nineteenth century a copper stained, blue variety of smithsonite (zinc carbonate, usually coloured white or grey) was found in the Malham mines of Lord Ribblesdale. So impressed was he by the beauty of the mineral that he had samples ground and polished into sequins which were sewn on to Lady Ribblesdale's ball gown. This caused a great sensation when she next appeared at a court function.

The list of allochromatic minerals is a lengthy one. It is best not to worry too much about colour for identification purposes when there are so many other good diagnostic features which can be used.

The colour of the streak made by scraping a mineral specimen on an unglazed porcelain plate may not be the same colour as that of the original mineral. The reason for this is because it reveals the colour of a thin layer or powdering of the mineral. Indeed, streak is often a more reliable means of identification than colour.

Lustre

Lustre is closely allied to colour in enhancing the aesthetic properties of a mineral. Like colour, it depends on the way the mineral reflects light. Reflectance, in turn, depends on opacity, transparency, impurities, structure and the presence of inclusions. A related feature is chatoyance, the ability of a crystal to reflect colours in the manner of shot silk, flashing with brilliance whenever light from a particular direction strikes it but appearing dull or differently coloured in other lights. In nature the reflection from a kingfisher's rump or a peacock's tail are unparalleled. The mineral world contains many similar delights of which chatoyant blue labrodorite-feldspar is a good example. This mineral is often found in laurvikite; a favourite decorative stone which is often used to face banks and public buildings.

Precious opals not only demonstrate the milky opalescence of common opal, but also have a deep-seated stress structure

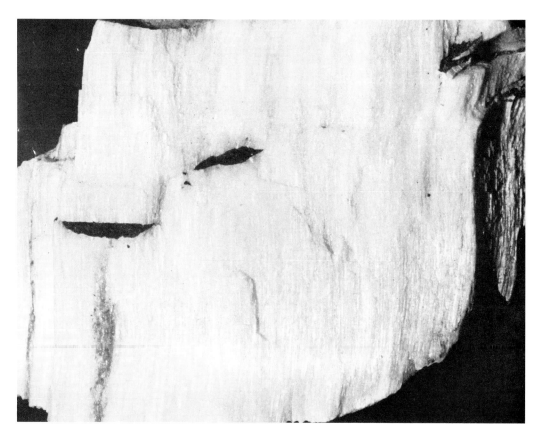

Fibrous gypsum is often called satin spar because of its lustrous sheen.

which displays all the colours of the rainbow in good specimens. Diamonds are said to have a brilliant or adamantine lustre which may be closely related in appearance to the vitreous (glassy) lustre of quartz and fluorspar. Other terms used to describe lustre are self-explanatory; they include earthy, pearly, metallic, silky and resinous.

Specific Gravity

The specific gravity (S.G.) or relative density of a mineral is an important diagnostic aid in the field when used in its simplest form by merely weighing a mineral sample in the hand. With experience it is relatively simple to gain a rough idea of a mineral's specific gravity or at least to be able to compare it with other nearby rocks and minerals of known specific gravity. To determine the specific gravity, or more accurately, weight per unit volume, it is necessary to weigh the sample first in air, then in water (S.G. of water is 1). A spring balance, with the specimen suspended on a length of fine fishing line, is quite adequate for rough weighing.

$$\text{Specific gravity} = \frac{\text{weight in air}}{\text{weight in air} - \text{weight in water}}$$

The specific gravity is stated as a number, without any units of weight or volume.

12

Other Properties

Several other physical properties of minerals may be useful to a lesser degree in identification. For instance, three minerals exhibit sufficient magnetic properties to be attracted to a magnet. These are pyrrhotite, ilmenite and magnetite. The latter is the most strongly magnetic of the three. Other minerals (calcite and, to a lesser extent, dolomite) are soluble in dilute acid. As it is inconvenient to carry bottles of acid in the field, solubility tests are best conducted in a laboratory with a wide range of dilute and concentrated acids. Minerals which are soluble in water may be identified by taste. Rock salt's identity is easily confirmed by this test. Ultraviolet light sources are also used occasionally by mineral collectors. The test is not conclusive but two fairly common minerals, apatite and witherite, fluoresce in ultraviolet light.

A knowledge of the crystal systems which determine crystal shapes is also necessary in the understanding of mineralogy. The arrangement of the crystal axes is the important feature. Cubic crystals have three axes of equal length which are mutually perpendicular. Hexagonal crystals have a long axis which is at right angles to three axes of equal length at 120 degrees to each other. This form of structure gives rise to the hexagonal prisms of quartz. Several minerals, typically calcite, form rhombohedral crystals in which there are three axes of equal length but with an angle other than 90 degrees between them. The faces of calcite rhombs are parallelogram shaped. Tetragonal crystals have two axes of equal length and one of greater length, all perpendicular to each other. Orthorhombic crystals are similar but all three axes are of different lengths. Crystals of the monoclinic system look like distorted crystals of

orthorhombic structure. They are similar except for the fact that one axis is not perpendicular to the planes of the other two. Finally, triclinic crystals have three axes of dissimilar lengths, neither of which is perpendicular to the other two.

The Vein Minerals

Quartz

Quartz, SiO_2 (silicon dioxide) may be found in almost any colour variety. The most common forms are colourless or iron stained. Quartz crystallises in short or long hexagonal prismatic crystals and is often found as glittering coatings in cavities. Quartz, or silica, is the most abundant rock-forming mineral and probably enjoys a similar status as a vein mineral, being found in company with many of the precious metals as a 'gangue' (spoil) mineral and also in association with base metals and other minerals. Quartz has a hardness of 7 and a specific gravity of 2.7.

Fluorite

Fluorite, CaF_2 (fluorspar), like quartz, often occurs massively (with no apparent crystalline structure). When crystalline, fluorite is cubic in shape. It may be colourless, purple, yellow, green or, occasionally, orange. Coloured specimens lose colour when heated. Fluorspar gained its name from the fluorescence of some specimens which appeared to be one colour when looked at, but a different colour when looked through. In Derbyshire the mineral was mined for decorative purposes. 'Blue John' vases were made from the blue-purple variety. Nowadays the mineral is important as a flux in metallurgy, and is

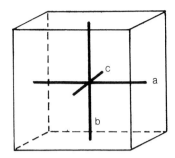

Cubic: a = b = c, mutually at 90°

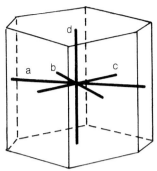

Hexagonal: a = b = c, mutually at 120° but at 90° to d

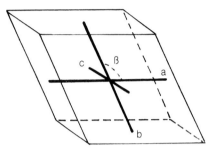

Rhombohedral: a = b = c,
mutually at < ß (not 90°)

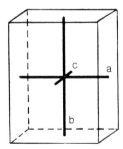

Orthorhombic: a, b, c = different lengths,
but mutually at 90°

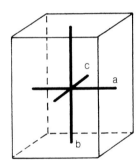

Tetragonal: a = c, axes mutually at 90°

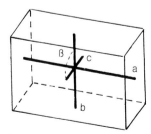

Triclinic: a, b, c = different lengths
at an angle other than 90° to each other

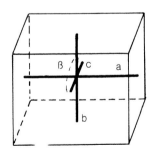

Monoclinic: a, b, c = different lengths,
a at 90° to b but b at < ß to c

Crystal systems.

14

Cubic crystals of fluorite.

used to produce hydrofluoric acid and in the manufacture of fluorine gas for aerosols and refrigerators. It is mined in Derbyshire and Durham. Its hardness is 4 and its specific gravity is 3.2.

Calcite

Calcite, $CaCO_3$ (calcium carbonate) is the main constituent of limestone. It also occurs in mineral veins, often in the form of rhomboid crystals which cleave into similarly shaped smaller versions of the original crystal. Calcite dissolves readily in dilute acid, producing bubbles of carbon dioxide. Most calcite is white or colourless, but it may be found in shades of brown, pink or blue. A particularly pure, colourless variety is called Iceland spar. An object viewed through Iceland spar is seen as a double image. Another carbonate, dolomite, has rhombohedral crystals with curved faces and a more pearly lustre than calcite. The hardness of calcite is 3, its specific gravity is 2.7.

Barite and Witherite

Barite, $BaSO_4$, is often known as 'heavy spar' for obvious reasons. Barite became the main economic product of many British lead mines, where it was found as a gangue mineral, when falling lead prices made lead mining uneconomic. It forms bladed and tabular white or colourless crystals of specific gravity 4.5 (much heavier than calcite). Its hardness is 3.5; slightly harder than calcite.

Usually associated with barite, witherite ($BaCO_3$) is a product of the action of

15

Hexagonal calcite crystals.

Bladed crystals of white barite or barytes.

carbonated water on barite, whereby the sulphate changes to carbonate. The procedure may be reversed by the reaction of sulphated water on witherite. Witherite has properties very similar to barite and its presence with barite was not discovered until 1784 when Dr William Withering published the first definitive account of the mineral found at Anglezarke in Lancashire. Eighty years previously Dr Charles Leigh, a Manchester physician, had described the diuretic and harmful effects of the mineral from Anglezarke on local inhabitants. At the time its chemical composition had not been discovered:

'The Neighbours thereabout will frequently take a Scruple at least of this in Fits of the Stone, in which it vomits, purges and works violently by Urine. There are some have been so daring as to venture to take a Dram of this, particularly James Barn's Wife and Child but alas! to their woeful Experience they found the Effects of it; for in about Nine Hours afterwards they both Expired ...'

As a result of its poisonous properties it was thought that the mineral would serve better as a rat poison.

Barite is not poisonous and has been used as a filler and as whitening agent in paints and paper. Nowadays barite is much in demand by the drilling industry which uses the mineral to maintain optimum specific gravity of the drilling mud when drilling for oil.

Although there are a few barite mines still operating, a large percentage of the supply is obtained by the reprocessing of spoil heaps at old lead mines. Barite is a common gangue mineral with lead, and is sometimes found with fluorspar.

Lead

Lead is obtained by smelting galena or lead sulphide (PbS). Galena is, perhaps, the most easily recognised of all the metallic minerals. It appears lead-grey and has a metallic lustre, tarnishing blue on exposure to the atmosphere. Unlike lead itself, the mineral is brittle, readily cleaving into miniature cubes of hardness from 2 to 3. Its specific gravity of 7.6 makes it the heaviest of the common minerals. When lead ore is smelted, or heated, to drive off its sulphur content, it yields about 87 per cent of its original weight as lead.

A lucrative feature of the British lead mining industry, which became almost defunct by the end of the nineteenth century (due to cheap foreign imports), was that most lead had a small silver content. About 6 oz of silver per ton of lead ore could be extracted economically. The extra revenue obtained augmented that from the sales of lead and, in some cases, paid for the mining processes so that the lead made a clear profit.

Zinc was often found in the same veins as lead ore. It belongs to the same suite of minerals, and most lead/barite mines would also produce zinc to a lesser or greater degree.

Sphalerite

Sphalerite, ZnS (zinc blend), has a resinous lustre and is usually brown, although ruby, yellow and black forms exist. When tested with hydrochloric acid, the distinctive smell of rotten eggs (hydrogen sulphide) is produced. Sphalerite crystallises into cubic, tetra or dodecahedral crystals, but often occurs in massive form. Its specific gravity ranges around 4, depending on its iron content. Its hardness is 3.5 to 4. When oxidised by carbonated ground waters, sphalerite changes to smithsonite or dry bone ore ($ZnCO_3$).

Iron Pyrite

One of the commonest sulphide minerals to be found as a vein deposit or as a replacement deposit in shales, mudstones, coal measures and many other situations, is iron pyrite (FeS_2), commonly known as 'fool's gold'. Its brassy yellow coloration has fooled many an inexperienced prospector. Recently, when the foundations for a hospital were excavated in a Scottish town, a vein of iron pyrite was revealed. Many of the local inhabitants attempted to stop the progress of building work in order to investigate the find which, to them, appeared to be a seam of gold. However, speculation ceased when the authorities ordered the hole to be filled so that building work could continue. There should, however, be little confusion between gold and fool's gold. The latter is hard (6.5) and brittle, whereas the former is soft and malleable. Pyrite tarnishes; real gold does not. Gold has a much higher specific gravity (19) than pyrite, which is relatively lightweight at 5. Pyrite is rarely used in iron making, but is useful in the production of sulphuric acid. Hematite, iron oxide, is the most important iron ore mineral.

Chalcopyrite

Chalcopyrite, copper iron sulphide ($CuFeS_2$), may be confused with pyrite. Like pyrite it is brittle, but it is less hard (4) and more golden in colour. It tarnishes readily into iridescent blues, purples and greens. Chalcopyrite is most frequently found as granular masses, often associated with quartz in medium temperature hydro-thermal veins. Both chalcopyrite and pyrite may contain small amounts of gold in solution. Perhaps the fools were not all that foolish after all!

When chalcopyrite is chemically attacked by carbonated water, or in the presence of limestone, green and blue coatings of copper carbonates – malachite and azurite – are formed. Both are important ore minerals of the oxidised zone of mineral veins, and large deposits of malachite have been mined commercially. At one time it was not realised that malachite and azurite could be smelted to produce copper and so they were discarded on spoil heaps.

Precious Metals

Even precious metals were at one time either ignored or thrown away. When the conquistadores (literally, searchers of gold) found nuggets of the white metal, platinum, in the streams of the New World, they realised that they had not found silver. Thinking that the metal was gold in an infantile, undeveloped condition they threw it back into the rivers to mature.

Gold may be found in Britain. In the Clogau district of Merioneth there exists an ancient gold mine which has provided gold for royal wedding rings since 1923. Gold placer (alluvial) deposits yield small flakes of gold to expert panners in the streams of the area. Helmsdale, in the Scottish Highlands, is also a good district for gold panning. That is not to say, however, that appreciable amounts of gold may be found, and there is little chance of a gold rush in Britain. Nevertheless, there are good chances of finding gold in commercial amounts in several places in Britain, and feasibility studies have taken place in Snowdonia and Argyll where gold is associated with other minerals.

One mineral, graphite, has declined in value but was once worth its weight in gold. Graphite (plumbago, or wad) is an allotrope (variety) of carbon, and is also related to diamond. More recently it has been used as a lubricant and as the 'lead' in pencils. In the past it was considered to be so important that it merited protection by Act of Parliament. The Act stated that plumbago was necessary for 'divers useful purposes in the casting of bomb-shells, round shot and cannon-balls'. It was a crime to enter the mines unlawfully. Nevertheless, there was an illicit trade in graphite which was sold unlawfully and smuggled across the fells by pack pony to Whitehaven and Maryport to be illegally exported. At the time of the Napoleonic Wars a permanent armed guard was kept at the entrance to the mine near Seathwaite in Borrowdale (Cumbria). All miners leaving the mine at the end of a shift were searched.

Graphite had strong medicinal properties and was taken to ease the 'pain of gravel, stone and strangury'. It was beaten into wine or ale and taken as required. Apparently, the mixture was a powerful diuretic and caused sweating and vomiting. If you were not ill, graphite was still useful; it was rubbed on swords, pistols and iron-work in general to prevent rust. Perhaps few other minerals have had such a diversity of uses.

Graphite – now used for pencil 'leads'.

Rock-forming Minerals

Feldspar

Apart from quartz, one of the most ubiquitous of minerals is feldspar, found as potassium aluminium silicate (orthoclase feldspar) or alternatively as sodium and calcium aluminium silicates. The pink or white crystals in granites are usually feldspar. Certain types of Cornish granite weather into soft clayey material. By using powerful jets of water, the decayed feldspars are washed out of the 'rotten' granites and collected. This is the basis of the china clay used in the porcelain industry.

Mica

The other silicate constituent of granite is mica, the black, white or colourless flakey mineral. Large crystals of white mica were worked commercially in Russia. This type of mica was appropriately named 'muscovite' mica. It is used as an electrical insulator and large flakes were mounted as 'windows' in slow combustion stoves and oil lamps. Glittering Christmas cards are dusted with small crystals of mica. The mineral is often used as a lubricant and in the manufacture of some types of rubber.

Garnet

Garnets are found in altered older rocks as, for instance, in the garnet-mica-schists of the Scottish Highlands. Good blood-red specimens of garnet may be used as semi-precious gem stones. Smaller, more crudely formed, brown garnets are used as abrasive powders. Garnets are silicates of a mixture of metals, including magnesium, aluminium, calcium and iron.

Asbestos

Asbestos is yet another useful silicate. It forms long fibrous crystals which may be spun and woven into asbestos materials renowned for their fire-resistant qualities. Unfortunately, many asbestos minerals act as a carcinogenic irritant and they are being replaced, wherever possible, by other, less harmful compounds. There is, as yet, no really suitable substitute for asbestos brake linings in motor vehicles.

Asbestos is a type of hornblende, a glassy green-black hydrous silicate of calcium-sodium - potassium - magnesium - iron - aluminium. Hornblende is one of the major rock-forming minerals, and is found in metamorphic and igneous rocks and in basalt lavas. Hornblende and some types of asbestos are classified as amphiboles.

Olivine and Serpentine

A silicate with a simpler structure, olivine is also green coloured but is heavier than hornblende. It is found in massive igneous rocks, such as gabbro, or in large non-crystalline amorphous masses with no particular shape. The crystalline variety of olivine is the semi-precious gemstone peridot. In South Africa diamonds are associated with peridotite rocks.

Olivine's close relative, serpentine (hydrated magnesium silicate), is an alteration product derived from either olivine or hornblende. Serpentine may be marbled green, white or red and has a soapy feel. Cornish serpentine is carved into ornaments and souvenirs for the tourist trade.

The rock-forming minerals crystallised from hot magma or were altered *in situ* by chemical changes and heat. Mineral veins were produced in a different way.

Formation of Mineral Veins

It is thought that sulphide ore mineral veins crystallised from a hot solution of mineral salts which ascended from great depths. Opinions about the origin of the water vary considerably. The main schools of thought argue that the water is either of atmospheric origin, and has percolated downwards, or is of deep-seated origin, and has migrated upwards. Perhaps the true case is a mixture of both.

The next question is how and where the metallic sulphides were dissolved in the water. As traces of metallic elements may be found in most rocks, it is argued by some geologists that ascending waters leached out metallic minerals from the parent rock and redeposited them in concentrated form within fissures and faults. An earlier theory suggested that the minerals were dissolved at great depths then rose to the surface regions as vapours and solutions. The first theory is somewhat discredited by the amazing bulk of rock that would have to be leached to give a sizeable sulphide vein.

There have been recent attempts to find the answer to this problem. It was reported in *Country-Side* (Autumn 1984) that Russian drilling teams had reached depths of 11km and had found mineralised solutions at those depths. So far below the surface, temperatures and pressures are so high that most metallic minerals dissolve and remain in solution until they reach cooler areas where they crystallise out of solution. By studying the assemblages of minerals from a vein, a geologist may hazard a fairly accurate guess at the temperatures which obtained at the time of formation.

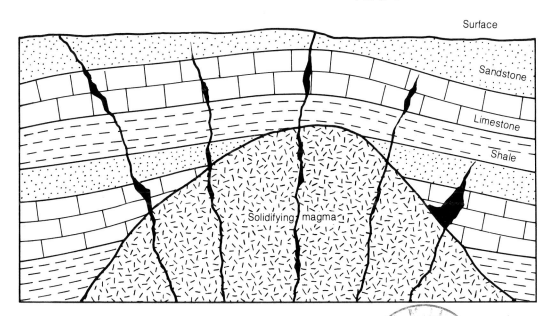

Injection of mineral veins from an igneous intrusive. Note how limestone is a good host rock and allows veins to spread.

The melting point of olivine is 1890°C; therefore the presence of olivine indicates high formation temperatures. Pyrrhotite (magnetic iron sulphide) has a melting point of over 1150°C. Furthermore, pyrite changes to pyrrhotite and sulphur above 685°C. So the presence of pyrrhotite in a vein indicates fairly high formation temperatures. Quartz crystallises from solution at much lower temperatures. If amethyst quartz or smoky quartz are heated above 260°C they lose colour. So the presence of coloured quartz indicates low formation temperatures. However, the converse does not apply and the presence of colourless quartz cannot be said to indicate high formation temperatures.

In some mining areas of Britain, particularly the north Pennine orefield near Alston, there is a distinct temperature zonation with copper, lead, zinc and fluorspar nearer to the central area where hot solutions were first injected into the country (host) rock. Further away, in the more peripheral areas of the mining field, amounts of zinc and copper begin to dwindle, there is less lead and fluorspar is replaced by barite. Finally, on the very edge of the mineralised area, only barite and calcite are found. By the time the hot solutions had reached such distances from the centre of mineralisation, all the other minerals with higher temperatures of formation had already crystallised from

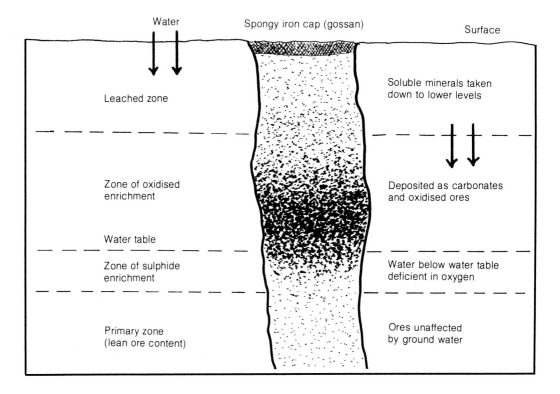

Secondary enrichment of a mineral vein.

solution. When searching for mineral veins geologists are encouraged by the presence of pyrrhotite, knowing that this is indicative of higher formation temperatures and that there is a higher probability of finding copper, nickel and associated metals.

Location of Veins

It is a characteristic of mineral veins that they are only found in certain localities. Why is it that some areas are so favoured and others are completely lacking in vein mineralisation? The answer is by no means definite, but it seems that mineralisation is influenced by the presence of granite, either at the surface or deep below. As we shall see in Chapter 3, granite is formed at depth. However, it contains large amounts of quartz and other minerals with low specific gravities. Even though the earth's crust appears to be solid, it is not, and granite masses formed at depth are constantly attempting to float upwards through the surrounding heavier rocks.

The presence of a large, underground granite mass may be detected by sensitive instruments which record variations in the earth's gravitational field. A low reading could indicate the possibility of underlying granite. A good place to search for minerals is often where a negative gravity anomaly has been recorded. As the masses of granite float upwards through the crust they push

Open cast working has exposed the outcrop of a barite vein at Strontian, Argyll.

at the overlying rock and cause fracturing and faulting. Below the granite masses there may be hot magma. Mineralising solutions percolate through fissures in the granites and then rise even further into surrounding rock. So mineral veins are formed as high temperature vein deposits in or near the granite (the tin and copper of Cornwall, for instance) and lower temperature deposits (fluorspar and lead) in surrounding country rock.

After the formation of a vein many millions of years may have elapsed before the miner or mineralogist finds it. Alterations have taken place and several minerals may be changed from their original state. Water is the main agent of change. In many dry regions of the world ore deposits have remained relatively unchanged since they were formed, but where there is high rainfall leaching occurs, leaving a cap of 'gossan' (a spongey, rust-coloured, leached area) from which all soluble minerals have been dissolved by the downward passage of water. Minerals from the gossan cap have been carried down from the leached zone to the oxidised zone (or 'zone of oxidised enrichment'). Here carbonates like azurite and malachite are found. Below the oxidised zone and beneath the water table, an area lacking in the oxidising ions of the ground water supply results in a zone of sulphide enrichment where secondary sulphides, such as chalcocite and bornite, are deposited. Lower still, in the unchanged area of the vein, ore concentration has not occurred and we enter into the primary zone which, lacking the water-concentrated deposits of the enriched zones, may be quite lean and often not worth the cost of mining.

Discovering a Vein

It is the gossan cap on a vein which often draws the prospector's attention to a mineral deposit. Sometimes the oxidised zone has been revealed by the scouring of glaciers, and colourful carbonates may be seen staining the rock. The amateur geologist is less fortunate. There are few possibilities of finding a previously undiscovered vein in highly populated areas of the world. However, there are many old mines and spoil heaps where, with the landowner's permission, it is possible to spend many hours looking for mineral

An almost vertical mineral vein in a fault fissure has been mined at several levels for lead and fluorspar. Appletreewick, North Yorkshire.

specimens. Excellent samples are still to be found, particularly on heaps that have not been reprocessed. Areas of particular note are the Devon and Cornwall tin/copper mining fields and the lead mining districts of Derbyshire, Yorkshire, Durham and Lanarkshire. Many beautiful examples of quartz, fluorspar and barite are still to be found. Rarer minerals require more diligent search.

Very little equipment is required. A geologist's hammer is standard equipment for many mineral seekers. Another useful piece of equipment is a small chisel of the type used by stonemasons. This weighs very little and, used with a hard stone or hammer as striker, applies force more accurately than would be possible with a geologist's hammer alone.

One of the biggest problems a mineral collector is likely to be faced with is storage of specimens. Therefore it is much better to collect good small specimens than large broken ones. A microscope eventually becomes an essential aid to identifying small specimens and for viewing crystal shapes. Although there are expensive mineralogical microscopes available, with filters for viewing thin rock sections in polarised light, the best type of microscope for an amateur geologist is a low power (10 to 20$\times$) stereo instrument. In the field a hand lens will suffice. The lens off a single lens reflex camera could be used as a magnifier if nothing else is available.

An Ordnance Survey map and a geological survey map of an area are essential so that grid references of good collecting localities may be noted for future reference. Geologists who are interested in the study of vein mineralisation should also take a hand compass so that the bearings of veins can be noted and their possible course followed to further exposures.

When visiting old mines remember that there are many deep shafts, some of them only covered by rotting boards or corroded ironwork. Extreme caution is required at every step. Always leave information of your whereabouts in case the worst happens. One Yorkshire lead miner was so keen to investigate an interesting prospect in a particular mine that he entered the mine in his best suit (he was on his way home from a wedding) without telling anyone where he was going. Whilst he was in the mine he was either the victim of an accident or suffered a heart attack. He was not found until almost a century later when

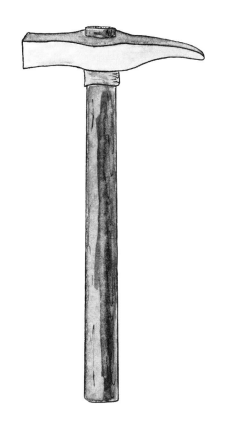

A geological hammer.

Spoil heaps at old mines are a mineral collector's paradise.

his body was discovered by a mine research group. This emphasises that mines should never be entered unless there is a back-up system on the surface or you are with experienced people. In any case, most of the interesting material has already been dug up and brought to the surface.

The Latin for an object 'dug up' is *fossilis*. George Bauer (Agricola) 1494–1555, was the first scientist to use the word, which he applied to minerals as well as to fossils – the subject of the next chapter.

2   The Fossil Record

Millions of years after the earth had condensed from a vapour cloud into a thinly crusted globe of molten matter, conditions were still too hot for the development of life. Slowly, as the earth cooled, oceans formed, an atmosphere of gases emanated from deep down and conditions were almost ready for the beginning of life. Try to imagine what earth was like so long ago. To us the atmosphere would have been hot, suffocating and poisonous, consisting of a high proportion of carbon dioxide, methane and sulphurous gases exuded from the vents of numerous volcanoes, the fumes and dust from which obliterated light from the sun. The humidity was probably high and violent storms were frequent.

These alien conditions were ideal for the nurturing and fostering of life. The high rainfall swilled mineral salts from the rocks into the seas which became a rich mineral soup of all the elements required for life. Add to this hydrocarbons from the atmosphere, the warmth of the climate and the electrical energy from storms, and the primeval earth was an ideal environment for the chance production of amino acids – the basis of life. Researchers have duplicated similar conditions in the laboratory and have succeeded in producing amino acids.

Even if life did not begin in this way; if the theories of Richter in 1865 (that life travels from planet to planet as minute spores carried along by the energy of light) were correct, conditions were still amenable for the development of life.

Whether life evolved or arrived, it is fairly certain that it began to develop either around the vents of hot geysers or in the warm oceans. We have no proof, because the first organisms – single cells – left no traces. As organisms developed in complexity they were probably minute, soft and jellyish, without skeletal structure and, therefore, incapable of leaving fossil remains.

What did the first organisms feed on? As there was no organic food, and cannibalism would not have been very productive when life was in its early stages, inorganic material must have provided sustenance. It is probable that the first forms of life were similar to the iron-sulphur bacteria of today, capable of deriving energy by oxidising and breaking down iron sulphides (iron pyrite) into oxides of iron and sulphur. Large deposits of bog iron ore (limonite) may be attributable to similar organisms, although no fossil trace of the original bacteria remains in these deposits.

At a very early stage in the development of life a division occurred. Perhaps feeble rays of light at last managed to penetrate the clouded skies. Some organisms developed the capacity to produce chlorophyll cells to utilise this new source of energy. (The coloration of some pre-fossil green rocks may be attributable to chlorophyll cells.) So began the great divide into animal and plant kingdoms, as some forms of life developed chlorophyll and others achieved the power of locomotion to move to new feeding areas or to capture other organisms for food.

As yet there were no animals or plants

with skeletal support, so the size attainable was limited by earth's gravity. Even under marine conditions there is a maximum size limit beyond which a living thing cannot grow without support. Furthermore, a large unprotected organism is vulnerable to many dangers. So, by a process of natural selection, the first animals and plants evolved with a more rigid structure. There was little fossil record until this time, but once living things had developed a sufficiently hard structure to leave a lasting imprint in surrounding strata an abundant record of the history of life had begun.

Life began at least 3·5 billion years ago. The earliest micro-fossils are found (in

Geological Time-scale

ERA	PERIOD	BEGINNING OF PERIOD (MILLIONS OF YEARS AGO)	MAJOR GEOLOGICAL EVENTS AND DEVELOPMENT OF LIFE
Cainozoic Quaternary	Holocene Pleistocene	0.012 2	Modern man Early man
Tertiary	Pliocene Miocene Oligocene Eocene Palaeocene	7 26 38 54 65	Mammals dominant Vulcanism in north-west Scotland. First primates
Mesozoic	Cretaceous Jurassic Triassic	135 195 225	Primitive mammals Age of reptiles
Upper Palaeozoic	Permian Carboniferous Devonian	280 345 395	First reptiles Hercynian Orogeny Pteridophytes & crinoids common Primitive fish
Lower Palaeozoic	Silurian Ordovician Cambrian	440 500 570	Caledonian Orogeny Graptolites and trilobites
Proterozoic and Archaeozoic		4,600	First life

South Africa) in microscopically thin sections of very early sedimentary rocks. They were single-celled marine organisms, bacteria without a nucleus or organelle cell parts, called prokaryotes. It was not until about 2 billion years ago that a new form of single-celled life – the eukaryote, with nucleus and organelles – began to evolve. Unlike prokaryotes, eukaryotes could use oxygen as a fuel for cellular processes. They were the forerunners of more evolved forms of life, the multicellular invertebrate organisms which began to appear in abundance about 600 million years ago.

Cambrian Period

By the beginning of the first period of the Palaeozoic era complex forms of life had developed. Perhaps the most characteristic Palaeozoic fossil is the trilobite, of which there are many forms. Modern day look-alikes of trilobites – sea slaters and wood lice – seem little changed in appearance although separated from the extinct trilobites by millions of years of evolution. During Cambrian times some species of trilobite became extinct whilst new forms evolved, so it is possible to date strata very accurately by the identification of a particular species of trilobite. Many species had developed eyes although some forms, such as agnostus, were blind. Early trilobites had long, spiny bodies with no facial suture, and their eyes were large and crescent shaped. Paradoxides, one of the larger species, evolved in Middle Cambrian times. Good specimens may reach a length of 50cm.

Ordovician trilobite, Angelina.

Trilobites were marine animals, so we may infer that wherever they are found was once covered by the Cambrian oceans. Although their ancestors have not been found, and were probably devoid of a hard shell, trilobites had well-armoured bodies.

It is likely, therefore, that the seas contained large amounts of dissolved calcium carbonate. This factor aided other organisms to form chitinous skeletons. Their fossils, too, have persisted.

Graptolites (from grapto, to write, and lithos, stone) are often called 'writing in the rocks', because their serrated skeletal remains resemble handwriting. They are found in abundance from late Cambrian times, often embedded between leaves of Cambrian slate. Early types were multi-branched and probably floated in the seas,

supported by a lung of air or gas. Monograptids developed later and continued through the Ordovician period.

Ordovician Period

Trilobites reached their peak of abundance during Ordovician times. Their fossils, with those of graptolites, are frequently found in shales, slates and sandstones which were deposited on the bed of the sea over much of southern Britain, Europe and most of China. The oceanic basin was ringed by mountains to the north, in Scotland, Scandinavia, Siberia and the Himalayas. Huge rivers flowed from the northern mountains depositing their coarse sediments in estuaries and more finely graded deposits far out on the ocean floor. Little change took place and the earth's crust went through a relatively stable period. The processes of evolution, however, continued at a rapid pace and the first crinoids appeared.

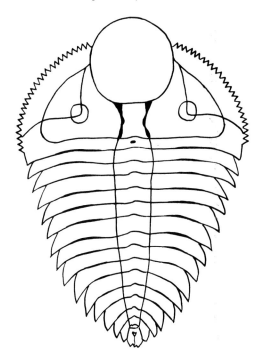

Trilobite – Staurocephalus clavifrons (Ordovician).

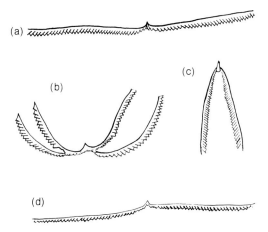

Ordovician graptolites: (a) Didymograptus hirundo; (b) Tetragraptus serra; (c) Didymograptus bifidus; (d) Didymograptus extensus.

Silurian Period

This final period of the Lower Palaeozoic era began in a similarly uneventful way, with much of Britain still under water. An event of great importance, however, took place in Silurian times: the first vertebrates began to evolve from worm-like creatures.

Significantly, the surviving relatives of the first vertebrates – the agnathids – are lampreys and hagfish, parasitic eel-shaped fish which attach themselves by sucker mouth parts to feed on their host's blood. The original agnathids had flattened, jawless, head parts with two upward-looking eyes set close together and an open mouth beneath a bony head plate. Fins were either rare or absent, but the body was protected by scales. Calcified vertebrae have not been found, so it may be presumed that the back was cartilaginous as are those of sharks. The side walls of the body were supported by gill rods which probably changed to ribs at the same time as evolutionary processes produced the calcified vertebrae found in later fish.

The next development in Lower Palaeozoic times was a fish with a jaw opening (gnathastomata). It is thought that the front pair of gills (there were ten gill openings in agnathids) had developed into a very primitive jaw. Well-preserved fossils demonstrate that the under-sized brain was attached to the upper jaw. As the jaw mechanism had developed from gills, movement was somewhat restricted.

Another variation – the placoderm fishes – had heavily armoured heads but only scantily protected tail parts. They are the only class of vertebrates to have become extinct. There are survivors from all other classes, even the most primitive. One of the longest surviving classes, apart from the agnathids, are the sharks and rays (Elas-mobranchs) of the Chondrichthyes. These are first found in Devonian strata.

Devonian Period

At the end of Silurian times there was a period of cataclysmic earth movements, known in Britain as the Caledonian Orogeny (mountain building period). Gigantic blocks of granite were formed at depth then thrust upwards over a period of millions of years. To the north of Britain massive earth movements thrust up the Caledonian Mountains.

The land that became Britain occupied a position just south of the equator. Much of the landmass was situated in hot equatorial regions. As thrusting, tilting and folding of older strata occurred, changing earlier rocks from shales and muds into hard slates, some of the creatures of the sea responded to the large areas of newly created land by colonising it. In all probability, the first fish to venture on to land were forced to develop rather quickly or face extinction when areas of water dried up. Unlike the mud skippers of today, the first amphibians had no choice. This is, however, a matter for speculation.

The earliest recorded fossil footprints have been found in Devonian strata, and the Upper Devonian rocks of Greenland contain the fossils of hybrid-like animals exhibiting both fish and amphibian characteristics – the Ichthyostega. Unlike fish they had a thicker, more supportive, backbone and lightweight tail structure. Their pectoral and pelvic fins had developed into primitive limbs, attached more strongly to the skeleton to provide extra thrust for walking.

Although Devonian times are noteworthy as the period when terrestrial vertebrates developed, another major event also occurred in the plant world. Vascular plants evolved with root systems and elongated cellular stems for pumping moisture and minerals to their leaves for the processes of photosynthesis and transpiration. The Pteridophytes evolved into the club mosses, ferns and horse-tails which have continued to thrive up to the present day. Although not as abundant as seed bearing plants nowadays, the Pteridophytes reached their acme in the coal forming era when species similar to club mosses and equisetums grew to heights of 30 metres or more.

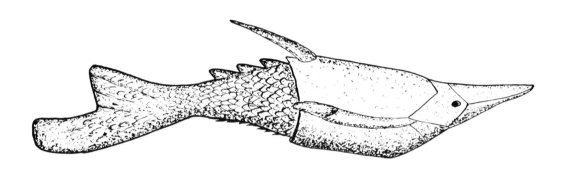

Pteraspis, a Devonian fish.

Ichthyostega, one of the first amphibians (Devonian).

Carboniferous Period

As the Lower Palaeozoic era ended, life on land was firmly established. Arid conditions were changing into a warm, damp climate. About 345 million years ago, a period began when the warm, tropical seas teemed with life to such an extent that a large proportion of the sedimentary rocks of the Carboniferous period were built up from the calcium carbonate skeletons of marine organisms. Crinoids, which date back to the Silurian period, reached maximum abundance in Carboniferous times.

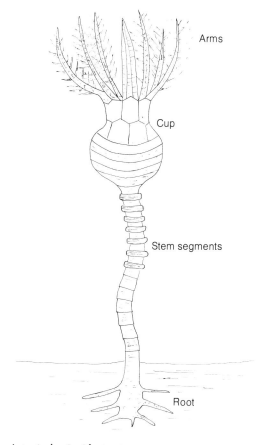

A typical crinoid structure.

Crinoids are wrongly called sea lilies, because they resemble plants. They were really food gathering animals. Crinoids lived in such abundance in warm mineral-rich seas that thick beds of rock are composed almost entirely of their remains. Unfortunately, most of these remains usually consist of broken stems; complete crinoids are rare. Crinoids were made up of an anchor to hold on to the sea-bed or to rocks, a stem and a cup with arms. The branched arms waved about in the water, collecting food particles which were passed down a cilia-lined groove to the mouth parts at their base. The hairy cilia vibrated and produced a downward current of water to transport food. Stem parts and the lower calyx were composed of ossicles (bony plates). In many strata these bony plates are the sole remains; calyxes are rare. Crinoids are often found in company with coral reef building organisms like lithostrotian.

Other common fossils of the Carboniferous period are brachiopods and cephalopods. Cephalopods – ancestors of the squid – were free swimming predators with armed tentacles. Goniatites were also particularly abundant at this time.

Goniatites were members of the Mollusca phylum, which includes cephalopods and nautilus with a coiled shell. They probably evolved from a creature like nautilus into the ammonoids of Devonian times, which in turn developed very quickly into goniatites, ceratites and ammonites (of the later Jurassic and Cretaceous periods). The important feature of goniatites is that they had a new type of suture line which was not slightly curved or straight, as were those of the nautiloid molluscs, but angular with forward and backward projections (called lobes and saddles). The siphon was not centrally placed, as in nautiloids, but off-centre on the outward side of the whorl.

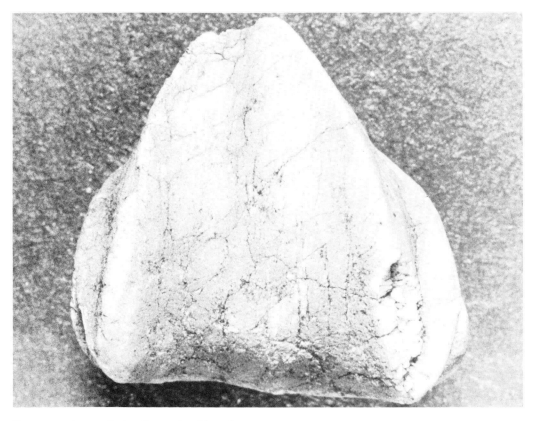

Pugnax – a Lower Carboniferous brachiopod.

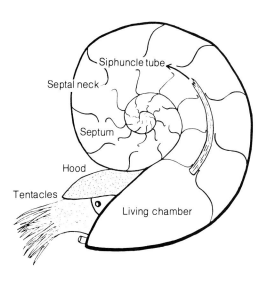

Siphuncle tube

Septal neck

Septum

Hood

Tentacles

Living chamber

Nautilus, a cephalopod. The shell in cross-section.

Goniatites were more compact and tightly coiled than nautiloids, rarely growing to more than 150cm in diameter. Small variations in structure are the clues by which different species of goniatite may be identified. They are extremely useful to geologists as a means of dating the Upper Carboniferous, which began with the Millstone Grit stage of deposition.

After a period of about fifty million years, during which most of Britain's Carboniferous Limestone areas were deposited, there was a change in conditions; not a quick change caused by massive earth movements and vulcanism, but a very slow uplifting of the land so that the seas gradually became shallower. The areas where Millstone Grit strata are found – the Pen-

Gastropod Pleurotomaria, Carboniferous.

nines and Southern Scotland – were estuarial regions where great rivers deposited deltas of coarse sediments eroded from the land areas to the north. As the deltas expanded and spread (rather like the Danube delta) parts of them became cut off from the sea and freshwater swamps and lagoons were formed.

The climate was hot and steamy; conditions were ideal for the growth of giant club mosses and equisetums. These grew to vast proportions then tumbled into the swamps. Fresh layers of vegetation grew upon their remains, but these also fell to the ground and suffered a similar fate. Growth was so rapid that the last layer had little time to decay before a new layer of vegetation collapsed on top of it. The result

was a lack of oxygen to fuel the processes of decay, so that thick layers of partially decomposed coal-forming vegetation soon built up.

Then, back came the seas to cover the swamps as the landmass slowly sank. Marine muds and shales were deposited on top of the swamps and goniatites re-established their domain for a short time. (Goniatites are common fossils in the grey-black shales above the partially formed coals). After a while more river delta sands and grits were washed over the muds and the whole process of delta building, swamp formation, coal-forming and sinking began again. The Coal Measures are typified by a cyclical succession. In each marine transgression, when mudstones and shales were

Mariopteris – a Coal Measure plant.

deposited, a new type of goniatite evolved. By studying their fossil remains it is possible to date and compare coal seams from different areas. Gastrioceras listeri, for instance, was often found in the Lancashire coal measures of the Bullion Mine marine band coals.

The swamps of the coal-forming period were teeming with many forms of life. Giant dragonflies, with a wingspan of two or three metres, hawked for smaller flies, cockroaches and other insects amongst the club mosses and horse-tails. In the swamps ganoid fish and brachiopods lived (including the common orbiculoides). Amphibians were, perhaps, the dominant form of life and Pholiderpeton, a fish-eating genus, was particularly common. Some amphibians had reptilian features and, although the true ancestors of the first reptiles are not known, it is suspected that they were descended from labyrinthodont amphibians, so-named because of their complicated tooth structure.

Faulting, folding and uplift of the Hercynian orogeny (earth movements) came at the end of the Carboniferous period. The Permian and Triassic periods of New Red Sandstone deposition began.

Permian Period

The Permian period marked the end of the Upper Paleozoic era. Magnesian limestones, which run in a southward curve from the coast of Durham, were formed. Some forms of marine life suffered a decline in their evolutionary careers. Corals and sea urchins grew scarcer and many

Coal balls – concretions formed around plant remains.

primitive species of shark became extinct. In the southern land areas giant forests of seed ferns grew; to the north plant life was more varied although, as yet, there were no flowering plants.

The first reptiles evolved from a branch of the labyrinthodont amphibians early in Permian times. This was fortuitous because reptiles had no need to return to water to breed. Their breeding cycle could take place on land because their eggs were protected by a tough, leathery skin. At this stage in our history conditions were changing to a hot and arid climate with few major inundations. Suitable breeding sites for amphibians were, therefore, scarcer than in Coal Measure times.

The first reptiles had several skeletal improvements on the amphibians. Their shoulder joints were more robust for walk

ing on land. They had broader, less flattened skulls than the amphibians. Dental structures were smaller and less complex than the labyrinthodonts. Reptilian skins were horny and more protective than amphibian hides. Permian reptiles had hides which resemble those of modern crocodiles.

The 'Age of Reptiles'

The Triassic period followed the Permian. During Triassic times reptiles were small, mostly land-dwelling types. As the Triassic New Red Sandstone times came to an end reptiles had developed into many forms, from quadrupeds to bipeds, marine and terrestrial types, small and giant and, eventually, flying reptiles.

Ginkgo leaves. Ginkgo, or maiden hair tree, is the sole survivor of the
Ginkgoales, Gymnosperms, which first developed in Triassic times.

Triassic times heralded the beginning of the Mesozoic era. Mesozoic is Greek for 'middle life', and the name was given because the fossils of the era are so different from those of the preceding era. Yet the animals and plants of the Mesozoic were very different from those that exist today. Many people have never heard of the term Mesozoic, but there are few who have not heard its more popular title – the 'Age of Reptiles'. When reptiles are mentioned in connection with the Mesozoic, pictures of dinosaurs immediately spring to mind. Dinosaur, from the Greek, means 'terrible lizard'.

Many smaller reptiles resembling lizards, crocodiles and turtles have escaped popular notice because they were insignificant compared with the terrible bulk and ferocity of some of the dinosaurs. Yet the smaller reptiles survived, while all the larger reptiles suddenly became extinct due to cataclysmic events that are, as yet, little understood.

Whilst at their acme of evolution, dinosaurs dominated all environments. In the sea were ichthyosaurs – long-jawed, dolphin shaped creatures armed with razor-sharp teeth which they used to capture fish and cephalopod molluscs. Ichthyosaurs were viviparous and retained fertilised eggs within their body until hatching. Another

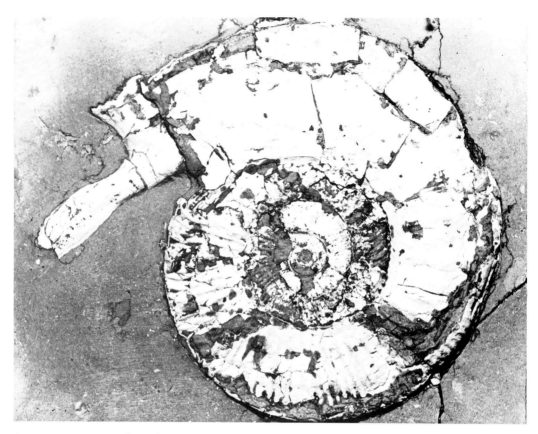

Ammonite comptonii, from the Jurassic period.

marine reptile, the plesiosaur, had limbs modified by evolution into powerful paddles for rapid propulsion through the sea. Plesiosaurs had huge seal-like bodies with long necks and reptilian heads. Adherents to the Loch Ness monster cult believe that the monster is of plesiosaurine type.

The sauropod reptiles, a group of quadruped herbivores, included the largest land animals ever to have existed. Both Brontosaurus and Diplodocus, of the late Jurassic period, could grow to lengths up to 20m and 27m respectively. Their relative, Brachiosaurus was about 22m in length, stood 6m to the shoulder and was extremely bulky. From evidence gained from examin-ation of fossil specimens, it has been calculated that Brachiosaurus could grow to weights in excess of 80 tonnes. What sort of muscles were required to move and support such bulks? It is likely that many of the heavier sauropods lived in swamps where the water gave some support to their titanic proportions. The motor control of such large bodies would present considerable problems if anything more than an inebriated stagger was required. Sauropod heads contained only a very small brain, but a much larger bundle of nerve tissue was situated in the enlarged spinal chord at the base of the tail. This may explain how co-ordination was achieved.

Myophorella, from the Jurassic period.

Biped herbivores, like the Iguanodon, had small upper limbs and long hind limbs. This enabled the reptile to rear up and graze on vegetation higher than other herbivores could reach.

Carnivorous dinosaurs were comparatively small, but very mobile. Tyrannosaurus was the largest carnivore, weighing close on 7 tonnes, with a length of 14 metres. Its head was about a metre long and armed with ferocious ripping teeth. The large, four-toed hind feet had three toes pointing forwards and one pointing behind. Each toe had sharp, curved talons for grasping food. The front limbs were small, but also sharply clawed.

Two important innovations during the Mesozoic era were flying reptiles and small mammal-like reptiles. In the flying pterodactyls ('finger-winged'), a wing membrane of leathery but supple skin was supported by an elongated fourth finger of the fore limbs and joined to the body at the side of the rib-cage. Flight was probably more of a gliding than of a flapping nature, as the breast bone structure of fossil specimens seems too small to have supported a strong musculature. Early pterodactyls had teeth in their elongated beak-like skulls. One of the later types – Pteranodon, of the late Cretaceous period – had no teeth, but a long, heavy beak which was counter-balanced by a similarly shaped protrusion on top of its skull. The Pteranodon had a wing span of 10 metres.

The second and to us the most important change in reptilian development was the evolution of reptiles with some mammalian characteristics. These were the therapsids, or mammal-like reptiles, which exhibited

Rhamphorhynchus, a Jurassic flying reptile.

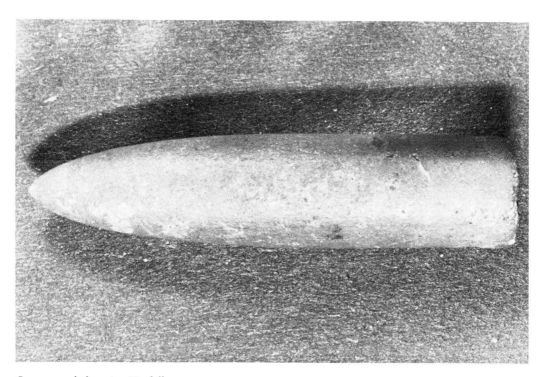

Cretaceous belemnite, Norfolk.

skeletal features such as differentiated teeth, limbs which were joined at the side of the body rather than beneath, and a more complicated skull structure. Tritylodon is thought to have been a link between reptiles and mammals. The first mammals of Jurassic times were of small stature (usually much smaller than a dog) and must have been timid, retiring creatures in a world dominated by gigantic reptilian carnivores. However, this state of affairs was short-lived. The age of dinosaurs came to a sudden end with the rapid extinction of all the larger reptiles at the end of Cretaceous times, 65 million years ago.

Extinction of Dinosaurs

Many theories have attempted to explain the nature of the catastrophe that brought about the end of the reptile age. Perhaps a sudden cooling of the climate, brought about by volcanic dust clouds obscuring the sun's rays, was responsible. Creatures with a large surface area would have been unable to retain sufficient body heat to remain active. Crocodiles and all modern reptiles are sluggish in cold conditions.

Another theory suggests that the giant carnivores became successful to the point of causing the near extinction of their herbivorous prey, thereby bringing about their own demise. A similar result would have been obtained by a lack of vegetation at the bottom of the food chain.

Recent research suggests reasons such as a change in the tilt of the earth's axis brought about by near collision with a wandering asteroid. A Russian scientist, Professor Neruchev, thinks that radiation from absorbed uranium compounds killed the dinosaurs. Furthermore, it is suggested that such 'biospheric suicides' take place every 30 to 40 million years with the mass extinction of many species. Successful mutations of other species may be caused by the same process. Apparently, some bacteria may gain more than half their dry weight by extracting uranium from water or the soil. One lake in the Kirghiz mountains has an unusually high number of algae species and ten times more uranium than is found in the oceans. When concentrations of uranium reach critical levels, the extinction of some species occurs. Many dinosaurs lived in shallow, warm water and, Professor Neruchev says, their remains have shown high concentrations of uranium. His theories do not explain how land-dwelling species became extinct, however. Personally, I suspect climatic changes were a major factor, and some recent research has shown that extinctions may have been triggered by the equivalent of a nuclear winter, brought about by soot clouds obscuring the sun after fires resulting from collision with a meteorite.

Cainozoic Era

After the extinction of the dinosaurs the era of recent life, the Cainozoic, began 65 million years ago. By the end of Cretaceous times there were many forms of life which resembled those of today. The first ancestor of birds – Archaeopteryx – had developed from reptiles back in the Jurassic period. By the beginning of the Cainozoic, in Tertiary times, birds were well advanced, hardly resembling their pterosaurian ancestors. They were well feathered on their body, wings and tail, and with this feathery insulation they were able to regulate body temperature. Also, their skull had greater capacity for a much more advanced nervous system than that of their reptilian ancestors. Many of the very primitive birds had solid, heavy bones, but most later types developed hollow bones which gave them a lighter structure to facilitate prolonged flight. Some of today's most primitive birds, the divers, first developed in the Cretaceous period. They have heavy bones and have retained this characteristic probably because flight is of secondary importance compared with their need to dive and stay under water with little effort. Unfortunately there are few fossil remains of birds with hollow bones; their fragility restricted preservation.

Plants

With the beginning of Tertiary times, the flowering plants (angiosperms) had become the most abundant and successful forms of plant life. By now the club mosses, horsetails and ferns were reduced in size and less common than conifers and angiosperms. The first ten million years of the Tertiary period were remarkable for the last major period of volcanic activity, particularly in Ireland and Scotland. The Tertiary lava flows of Mull contain the imprints of tree trunks which have been replaced by basalt. In quiet periods between volcanic outbursts, thin fossil soils containing a rich variety of flowering plants accumulated. From fossil evidence found in London clay we know that the climate was sub-tropical; Malayan palms, oriental species of ginkgo and oak flourished.

Mammals

Mammals were able to take over as the dominant type of animal life after the extinction of the dinosaurs. They began to expand their range and develop into forms to fill all the vacant ecological niches that had been vacated by the dinosaurs. At the end of the Jurassic period it had been doubtful whether any mammals would survive; only one group, the pantothenes, remained. Before the end of the Cretaceous, they had diversified into marsupial and placental mammals. Marsupials became the dominant mammalian type for a short time, but their position was soon usurped by the sheer diversity of placental mammals.

By the beginning of the Cainozoic, placental mammals had adapted to life in a variety of environments. Some had taken to the sea and others to the air, but the majority were thriving on land. Placental mammals never reached Australia, which had already become separated from the main landmass. There, marsupials developed without competition. A similar situation applied in South America, until North and South America were joined together by the land bridge of Panama. Since that time marsupials there have regressed in dominance.

Australasian marsupials developed to exploit almost all the available habitats. Specialised marsupial species closely resembled their placental counterparts. There were cat-like carnivores – the Borhyaena and Prothylacinus. A large tigerish marsupial, the Thylacosmilus, resembled the placental sabre-toothed tiger and even had similarly enlarged upper canines. Although these three animals are found today only as fossils, there is a possibility that a close relative may still survive in the flesh. The thylacine or Tasmanian tiger actually resembles a wolf more than a tiger. It was thought to have become extinct in the 1930s as a result of trapping and poisoning incurred by its predation on sheep. Recent reports of sightings indicate that a few thylacines may have survived to the present day. From the fossil record, thylacines were widespread in

Australia and New Guinea up to about 3,000 years ago, when the dingo arrived in Australia.

Marsupials came second in the evolutionary contest to produce the dominant animal because no primate-like marsupial has ever existed. The placentals produced the primates, which are of some importance in the evolution of life because they are the ancestors of man. Fossil primates date back to the Palaeocene times of the Tertiary period, 65 million years ago. From them came lemurs, monkeys and apes. Apes diverged into gorillas, chimpanzees and orang-utans (the pongids) and into the hominid group which embraces man and his ancestors.

Man

It is unclear when and where the first man-like creature developed. As investigations proceed in various parts of the world, and new contenders emerge for the status of 'missing link', the picture seems to become more clouded, the information more confused. How do we judge whether an animal is hominid or not? The ability to use tools and to walk upright may be essential characteristics, but there is no reason why some form of culture could not have developed in creatures who travelled on all fours. A degree of consensus has been arrived at, and it is now recognised that *Australopithecus africanus* is the first recognisable fossil ancestor dating back to 2 million years ago, at the end of Tertiary times. The Olduvai Gorge in Tanzania has yielded a stratified archaeological museum of man's progress, including remains of implements and tools made by *Homo habilis* of about 1.85 million years ago.

After *Homo habilis* there is a possible red herring in *Paranthropus robustus*, a heavy

browed primitive creature who may have evolved at the same time but died out about 900,000 years ago as a result of his inability to compete with his more intelligent cousin.

Homo erectus, of the early Palaeolithic, has also been found in the Olduvai Gorge, but at a higher level approximately 1.1 million years ago. Anatomically related hominids of a similar age have been found in various locations, often with remains that prove they used primitive stone tools and fire. *Homo erectus* evolved into Neanderthal man, the ancestor of *Homo sapiens*, who demonstrated a much more developed sense of dexterity in the use of stone tools and in the wall paintings executed in caves at Altamira and Lascaux. The age of modern man began 10,000 years ago, and the fossil record, which began with graptolites, no longer had the monopoly of history when man invented writing.

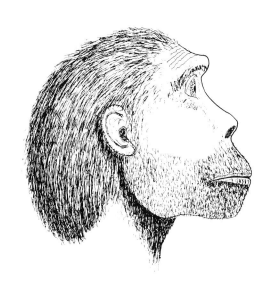

Pithecanthropus, Tanzania.

Collecting Fossils

Collecting fossils is very much like collecting anything else. You may decide to collect everything from anywhere or, better still, to specialise in a particular period or to concentrate your efforts on only a few groups of fossils. Probably the best way to begin is to pick a familiar area in which the geological succession is fairly limited. Buy local geology maps at a scale of 1:50,000 (or the old 1-inch type). These are available from the British Geological Survey (formerly the Institute of Geological Sciences).

Geological maps of some areas are available as 'special sheets' at a larger scale. These often show good locations for fossil collecting. BGS also publish memoirs to the geological sheets at 1:50,000. These invaluable books describe areas in detail and give grid references where fossils and other points of interest are located. However, to rely too much on the information provided by others takes much of the fun out of geological field work. It is often better to go into the field without the aid of maps in order to try and identify strata from their fossil content and succession.

Many important fossil bearing strata are unearthed by the activities of quarrying companies. It is usually impossible to enter working quarries for reasons of safety, although permission may sometimes be obtained at weekends. Some of the best quarries, therefore, are disused and abandoned, but unfortunately these are often

A fossil of the future – the structure of a modern coral.

used for refuse disposal and are filled in and levelled off. I have seen several good fossil localities disappear in this way. The availability of fossil sites, therefore, is constantly changing. Hence the need to find new ones, perhaps where landslips have revealed a new rock face or streams have cut new strata.

There are some geologists who cause more erosion in a few minutes than the natural processes achieve in thousands of years. Never begin your own mining or quarrying operations where a cliff may be made unstable or a farmer's good pasture ruined by your digging. To avoid becoming a public nuisance, the most suitable way to collect is to remove only those fossils which are already revealed by natural erosion. A small amount of chipping with a hammer and chisel is acceptable, but be patient and careful and try to remove the fossil in one piece. If it is already fragmented, attempt to collect all the pieces. They can be stuck together later.

If you are lucky enough to find a large fossil which looks like an important find and is only partly protruding from a rock face, seek expert help rather than hacking away at it and spoiling it. Many important finds have been made by people who have had the forbearance to seek the help of a local expert from a museum, who can arrange for a careful excavation and proper identification.

To avoid frustration and wasted effort, take careful notes at the time of collection: when, where and in what stratum the specimen was found. Always carry an inexhaustible supply of polythene bags for individual packing to avoid confusion with other specimens.

Some fossils may be covered in an adhering patina of clay or stone. By careful use of a small drill of the type used to drill printed circuit boards (obtainable from electronics stores), it is possible to clean away encrusting debris. Siliceous fossils encrusted with limy deposits may be cleaned in acids which will dissolve the calcium carbonate.

Sometimes only the imprints of fossils are found, the remainder having been dissolved by ground water or decayed by the bacteria of millions of years ago. Use plaster of Paris to get a good impression. Make sure that the original surface is sprayed with washing-up detergent to facilitate removal of the hardened cast.

If you wish to look at the internal structure of a fossil like a goniatite, do not try to chip it in two. The result will probably be a hopeless mess. It is far better to grind the fossil down using a thick piece of glass as a base-plate. On this sprinkle some water and coarse carborundum powder (120 grade). Use progressively finer grades to polish the specimen after you have finished coarse grinding. Make sure the coarser particles are washed away before changing to finer grades. Finally, polish with cerium oxide powder to obtain a mirror finish, then protect the finish with a layer of clear lacquer.

3   Rocks

For the sake of scientific convenience, geologists group the rocks of the earth's crust into three types – igneous, sedimentary and metamorphic. It is often difficult to decide which of these categories applies to a rock; indeed, it is possible to find rocks of similar chemical composition within all three groups. The deciding factor in the classification of a rock is, therefore, its mode of formation rather than its chemical composition. Igneous rocks consolidated from magma. They are often referred to as volcanic or eruptive rocks, but the defi-

nition 'igneous' encompasses all modes of formation from magma. Sedimentary rocks are the products of weathering and deposition of older rocks. Metamorphic rocks have been changed from their primary state (igneous or sedimentary) by heat or pressure.

Igneous Rocks

All igneous rocks ascended in a fluid state and cooled from hot magma into a crystalline structure. It has been pointed out that of all the igneous rocks of the earth's crust, 90 per cent are either granite or basalt. This is an over-simplification of the true state of affairs, as there are many varieties of granite and basalt with intermediate types. Granite contains large amounts of silica, whereas basalt has a low silica content. By analysis of the amount of silica in an igneous rock it is possible to determine whether it is acid, intermediate, basic or ultrabasic. Free quartz is found only in acid rocks. Intermediate rocks contain silicates in the form of feldspars. In basic rocks feldspars are sparse. Ultrabasic rocks are identified by the absence of alkali feldspars. Even this classification is too simplistic to include some igneous rocks which transgress these artificial boundaries.

In the table some of the commoner igneous rocks are classified according to composition and granular structure. Although the table does not cover all possibilities, it is a help in classification. For instance, an unknown coarse grained igneous

Granite at Illgill Head, Cumbria.

Classification of Igneous Rocks

	Crystalline Texture	ACID	INTERMEDIATE		BASIC	ULTRABASIC
			Feldspar Orthoclase ←→ Plagioclase			
Extrusive ↑ (*Usual Occurrence*) ↓ Intrusive	Fine	Rhyolite	Trachyte	Andesite	Basalt	Ultrabasic lavas
	Medium	Microgranite	Microsyenite	Microdiorite	Dolerite	Peridotite porphyry
	Coarse	Granite	Syenite	Diorite	Gabbro	Peridotite
		2.5 ←————— Specific Gravity —————→ 3+				

rock sample could only belong to either the granite, syenite-diorite or gabbro families. If it contained no free quartz, very little alkali feldspar (orthoclase), but with soda lime feldspar (plagioclase) predominating, it would be a diorite. Acid igneous rocks usually contain orthoclase feldspar, quartz and mica. If composition tends towards basic, augite or hornblende may be present. In intermediate rocks hornblende and feldspars dominate, but slightly acid intermediate specimens still contain small amounts of mica or quartz. At the basic end of the scale augite begins to dominate, in association with plagioclase feldspar and some olivine. Finally, the ultrabasic rocks may contain no feldspar at all. They tend to be almost entirely composed of a single mineral – typically olivine – and are often called the green rocks. The specific gravity of ultrabasic rocks is usually 3 or more. Compared with acid rocks they are much heavier. In the field a simple hand-weighing test is often a useful aid to identification.

Grain size is usually determined by the rate of cooling from the magmatic state. The rate of cooling is, in turn, affected by the conditions of formation. Rhyolite, trachyte, andesite and basalt lavas are often formed as a result of volcanic (extrusive) activity or by intrusion as thin sheets (dykes and sills) into surrounding rocks. In either instance they are fine grained because cooling was rapid and there was no time for the seeding and growth of large crystals. Coarse grained rocks were intruded in large masses at depth, where they were insulated from the effects of rapid cooling by surrounding rocks. Therefore, large crystals had ample time to form.

Intrusive Rocks

All granites are coarse grained. Their composition complies roughly with the formula above, although individual types may vary considerably in appearance. Some specimens of Cornish granite contain very large crystals of white feldspar. Shap granite from Westmorland is a favourite for monu-

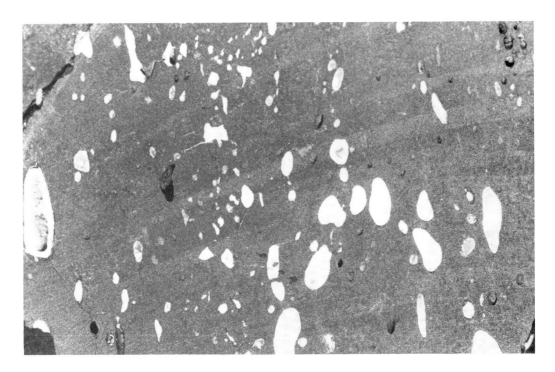

Vesicles of zeolite in basalt.

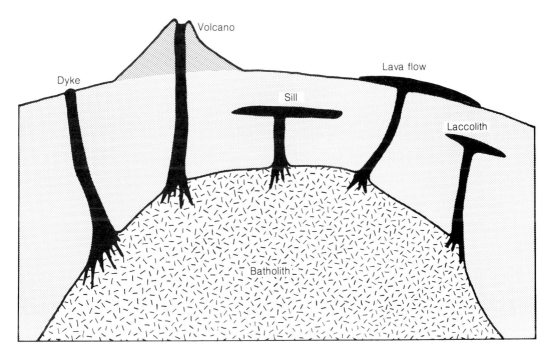

Igneous intrusives and extrusives.

mental masonry and for facing buildings because of its large, pink feldspar crystals. Aberdeen granite is a hard, cold stone with steely-grey crystals of feldspar and a glint of mica. Although some granites are hard and difficult to work, others tend to crumble. In some Cornish granites the feldspars have rotted in the process of kaolinisation which produces the invaluable China clay. Some Lake District granites have a similar tendency to disintegrate.

Syenites contain scarcely any free quartz. Many are handsome rocks and are used in similar ways to granite. One popular type of syenite from Norway, laurvikite, graces many of our large public buildings and shops. When polished, its mottled grey appearance is enhanced by the shimmering blue chatoyance of its plagioclase crystals.

Gabbros are composed of tightly packed crystals of plagioclase feldspar and dark-coloured augite. When compared with similar sized specimens of granite they weigh heavily in the hand, as they are completely lacking in lightweight, free quartz. They are the product of basic intrusions, formed during periods of volcanic activity. In Scotland there are Tertiary gabbros from the period of volcanic activity in the north-west.

Intrusions

Large masses of igneous rocks which cooled beneath the earth's surface are called batholiths. Even when recent erosion has not uncovered them at the surface, it is possible to detect batholiths by using sensitive gravity recording equipment. Two hidden masses of granite were first suspected by the detection of negative gravity anomalies centred in Weardale and in Wensleydale. By drilling through overlying strata, as near as possible to the centre of each anomaly, the crests of both masses were located and the presence of granite proved. It would seem that much of the northern Pennines and Yorkshire Dales are underlain by granite at no great depth. The presence of smaller intrusions would be more difficult, but still possible, to detect.

Many smaller intrusions – laccoliths, phacoliths, lopoliths and sills – may not be very extensive vertically but often extend great distances laterally. They have often been fed with magma from below by a dyke or several dykes. Laccoliths are intrusions of domed appearance with a flattened under-surface. Their emplacement had the effect of forcing overlying strata upwards. Laccoliths and sills may easily be confused, but only if the laccolith is of fairly uniform thickness. Conversely, lopoliths are a similar type of intrusion, but with upper surfaces concave and under-surfaces convex, forming a saucer or basin shape, as in the Sudbury Basin of Canada – an important area for the production of nickel. Phacoliths are masses of lava which were injected into folded strata where there were gaps or weaknesses in the anticlinal and synclinal ridges.

Sills are found at a particular horizon between two strata, only changing horizons when diverted by a fault or similar obstruction. By definition, a sill lies along the bedding planes of host rock. The Great Whin Sill, whose northern outcrops in the wild Border country of Northumbria provided the foundations for Hadrian's Wall, is a splendid example of a sill.

Dykes are of similar occurrence and structure, but cross strata rather than lying parallel with bedding planes. Dykes are often associated with eruptive volcanic activity, having been intruded into surrounding rock at the same time as volcanic eruptions, or a short time afterwards.

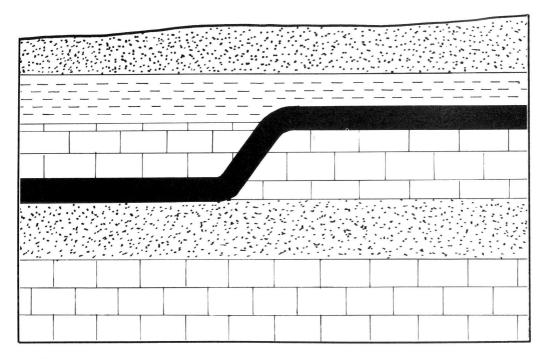

A sill changing horizon.

Although some dykes were injected over great distances (several dykes in northern England point towards and are thought to be associated with Tertiary vulcanism on Mull) and may extend for many miles in a straight line, others are shaped into circular formations surrounding the cones of extinct volcanoes.

Vulcanism

Two excellent areas for the study of vulcanism are the Ardnamurchan peninsula and the Island of Mull in north-west Scotland; both of these were volcanic centres in Tertiary times. Similar events have taken place more recently in some of the Pacific volcanic areas.

The first stage of volcanic activity in Mull, Ardnamurchan and Skye was caldera formation. A large, seething cauldron of magma collected within the crust, from which successive sheets of fluid basalt lavas poured. Each eruption blanketed many miles of surrounding land with a thick layer of consolidated lava. Basaltic lavas are extremely fluid; capable of flowing great distances from source in a matter of hours.

As the original reservoir of lava became depleted, sinking of the central caldera occurred, causing the formation of a circular depression several miles across. Then, more basalt lavas were extruded, mainly from the edges of the caldera, and still greater thicknesses of basalt accumulated over the landscape. The next stage was probably the extrusion of viscous acid lavas which, because of their high viscosity, were incap-

Dolerite in the Whin Sill at Cauldron Snout, Durham.

able of travelling very far – the result being the build-up of the volcanic cones of central vent volcanoes. Such cone volcanoes tended to be self-blocking, as activity temporarily declined, until fresh eruptions of renewed vigour removed the consolidated lavas of previous eruptions. The resultant explosions blasted away the top of the old cone with the force of an atomic bomb, creating vast clouds of ash, a bombardment of boulders and clouds of noxious sulphurous gases.

So the processes alternated between basic and acid extrusions, each basaltic flow heating up the surrounding rock sufficiently to allow the passage of viscous acid lavas to the surface where volcanic cones sprouted at the edge of the calderas. Eventually, when eruptive energy decreased, magma was unable to reach the surface and the next stages of activity were marked by the deep intrusion of gabbro. One such mass of gabbro forms the mountain of Ben Buie on Mull.

The history of a period of vulcanism may be studied wherever erosion has produced good exposures of successive lava flows and intrusions. Tuffs (consolidated ashes) and agglomerates (the debris of cone-shattering explosions) mark the beginning of renewed periods of activity, after which rhyolites were extruded. These alternated with basaltic lava sheets from the lava-plateau building stages. Each volcanic area has its own sequence, but the basic pattern remains the same. Tertiary volcanics are not only found

Basalt lava flows, Canna, Hebrides.

These large boulders in agglomerate are a result of Tertiary vulcanism at Canna, Hebrides.

in the north-west of Scotland, but also in Ireland (Giant's Causeway), Iceland and Greenland. The whole area is referred to as the North Atlantic (or Thulean) Igneous Province.

Ring dykes and cone sheets are a further complication in the study of vulcanism. Cone sheets, in an idealised cross-section, resemble the top of an ice-cream cone in shape. Where they have been exposed at the surface by erosion, they appear as concentric rings radiating outwards from a hidden block of consolidated igneous rock – their original feeder-reservoir. Conversely, ring dykes form an inverted cone, hading outwards from the centre of the ring. Cone sheet lavas tend to be medium to finely crystalline, typically felsite, acid

stages of vulcanism. They are of insignificant width when compared with ring dykes which may be a mile wide and, as a consequence of slower cooling, coarsely crystalline.

Origins of Igneous Rocks

It was once supposed that all igneous activity was triggered by the slow cooling and wrinkling of the earth's surface and the exploitation of consequent structural weaknesses by molten material rising from the cone. Recent research indicates that earthquakes and igneous activity are very closely related. A good atlas usually contains maps delineating areas of the world that are 'high risk' earthquake zones. Fortunately these

53

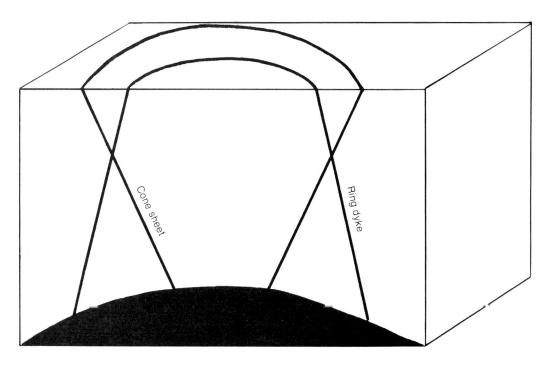

Block diagram of cone sheet and ring dyke formation around a hidden magma reservoir.

areas are only narrow seismic zones surrounded by relatively safe, medium and low risk zones. (The British Isles are situated in a medium risk zone.) The earth's surface is divided into a number of aseismic plates (crustal areas where earthquakes are uncommon). It is within the bands where the plates meet and grate together that there is a high risk of earthquakes and the possibility of volcanic activity.

At one type of plate boundary – oceanic ridge boundaries – the plates are in the process of moving apart. At the same time, basalt lavas are injected from below to plug the gap and enlarge the plates. When basalt lava is extruded directly into the sea a formation called pillow lava is produced as the lava quickly cools into sack-shaped lumps.

When two plates move apart it follows that, elsewhere, there must be a head-on collision. This type of junction of aseismic plates is called a destructive margin. But what happens to the surplus debris at the point of collision? Unless there is considerable upthrust of the land each time there is an earthquake, it is probable that a large amount of material from one of the colliding plates passes under the opposing plate and is re-absorbed into the mantle after being melted by the heat of friction at the point where the two plates collide. This process is constantly taking place along the west coast of South America where earthquakes and volcanic eruptions occur along the Andean chain. Where two oceanic plates meet destructively, a group of volcanic islands may be the result of increased activity. The Pacific islands of the Philippines have been produced in this way.

Igneous activity is always taking place, usually unnoticed and at great depth, but occasionally in spectacular eruptions like the Mount St Helens volcano in America. New islands, like Surtsey off the coast of Iceland, rise up as if by magic. Some disappear just as quickly as they were formed, victims of isostatic adjustment – slumping back into the ground that gave birth to them. Most igneous rocks have a longer life-span, even though the processes of erosion constantly wear away at them, producing the raw material for sedimentary rocks.

Sedimentary Rocks

Sedimentary deposits are sub-divided into two main types, the clastic and non-clastic. Clastic strata are the mechanically redeposited and consolidated remains of eroded older rocks. Non-clastic strata may be chemically or organically deposited.

As already noted, it is quite possible to find a sedimentary rock with a similar composition to an igneous rock. How would we differentiate between the two? A general rule is that igneous rocks fit together like a jigsaw puzzle; all the crystals interlock and hold together. Sedimentary rocks have been deposited in a random fashion and the spaces between the grains of mineral are usually occupied by a material which acts as a cement. The grains of mineral are often worn or rounded, lacking the crystalline angularity of igneous rock particles. Particle size is the most useful aid to classification.

Clastic Rocks

The coarsest grained sedimentary rocks are conglomerates and breccias. Even to the unpractised eye these rudaceous, coarse grained, rocks may be instantly classified into breccias (with angular rock fragments) and conglomerates (formed of rounded pebbles). The shape of the rock fragments tells us something about the distance they were transported from their eroded parent rock. Worn, rounded conglomerate pebbles may have been carried great distances from their original situation by water. Breccias are unworn, and were therefore deposited almost *in situ*. Both breccias and conglomerates are held together by a matrix which may be formed of the original sand or clay material that was deposited with the pebbles or, in some cases, a calcareous matrix that was deposited, from solution, within the spaces between pebbles.

By studying the fragments within rudaceous rocks it is possible to discover their original source and estimate the direction of flow of the rivers that carried them and the lie of the land over which those rivers flowed. The boulder clays of the ice ages are, as yet, unconsolidated; they will form the rudaceous rocks of the future. Boulder clays contain a mixture of both rounded and angular fragments picked up at various points by the slowly moving ice sheets and glaciers that, eventually, redeposited them. Such fragments provide an accurate record of the routes travelled by the ice.

If the particle size of a rock is below 2mm in diameter it is no longer classified as rudaceous. Sedimentary rocks with grains of $\frac{1}{16}$mm to 2mm diameter are arenaceous (or sandstone). Within these boundaries are found all types of sandstone from very coarse to very fine. The fineness of particles in a sandstone is a clue to the extent to which its grains have been transported. If conglomerates were formed on the flood plains of fast flowing rivers, then sandstones were deposited either further out in the river estuaries, carried to beaches, or deposited on the flood plains of slow flowing rivers that lacked the vigour to carry their sediments to the sea.

Most sandstones are formed almost exclusively of quartz. In the prolonged process of erosion and transportation downstream, many of the softer or more soluble minerals disintegrated, leaving the harder quartz as the sole survivor. However, many sandstones contain fragments of other minerals. The fissile flagstones which provide roofing tiles and paving stones split easily along their bedding planes because occasional calm water conditions at the time of deposition allowed fine platelets of mica to fall to the bottom and interfoliate between the sands in paper-thin layers. Red sandstones of Devonian and Permian times are stained red by iron compounds. The sands of Alum Bay, Isle of Wight, are a multi-coloured example of mineral staining. Jurassic glauconite greensands owe their striking coloration to minute crystals of glauconite – iron compounds. Some arkose sandstones – formed near source from the decay of granites – contain large quantities of feldspar.

Beds of sandstone are not produced merely by consolidation of sand deposits.

Millstone grit – a durable, coarse sedimentary rock deposited in Carboniferous times.

The pressure of overlying deposits is not sufficient to produce hard, durable rock without the addition of some form of cement. Interstitial water, in the pores between grains, may carry silica or calcareous compounds in solution. By complex chemical processes, which result in precipitation of these compounds, the pores are filled and the sand grains cemented together. There are many cementing agents; calcite is the most common, but substances like witherite (barium carbonate) are occasionally encountered.

Not all sandstones fit into a particular category of grain size. The greywacke sandstones are a complex mixture of particles, containing quartz, feldspar and even small fragments of rock in a matrix of smaller particles of the same minerals; to which may be added chlorite, illite, mica, dolomite, pyrite and microscopic epidotes. Their predominant colour is grey-blue. There is no satisfactory explanation for their mixed particle size, which contradicts all the standard theories of depositional grading. However, the consensus of opinion suggests that the particles were deposited by strong currents.

Siltstones, mudstones, shales and clay-stones – the argillaceous deposits – are the finest grained clastic sedimentary rocks. Theoretically, there are two extreme types. One is composed of rock flour, the finest product of mechanical weathering of the decomposed rock-forming mineral crystals. The other is a result of the chemical weathering of older rocks.

Rock flour consists of the ubiquitous components of igneous rocks ground down and abraded by water, ice or wind transportation. Quartz, feldspar and mica predominate over small particles of rarities like tourmaline and rutile. The chemical composition of the chemically weathered

clay materials is so complex that generalisation is the best policy. It is sufficient to point out that clay materials may be regarded as silica with oxides of aluminium and, to a lesser degree, oxides of iron, calcium, sodium, magnesium and potassium, with water in a hydrated silicate compound. Most clays have a layered structure which gives them their characteristic greasy feel.

Clays have many uses other than those with which they are usually associated. Not only are they used for pottery and brick making, but there are also clays from which bauxite is extracted for aluminium smelting. Seat-earth clays, from the base of coal seams, have been exploited for their refrac-

Horizontally bedded Yoredale series at Hardraw, North Yorkshire, consist of alternating beds of shales, limestones and sandstones.

tory qualities and are used to make fire-bricks and furnace linings. A good fire clay should be able to withstand temperatures of more than 1600°C without disintegration. Other clays have been used as constituents of fillings in chocolates. Alum clays and shales have been worked in various parts of Britain for alum, which was used in the dyeing industry.

Non-clastic Rocks

Often called 'chemical' rocks, the non-clastic rocks are products of chemical weathering and precipitation. The most abundant are the calcareous deposits – the limestones and chalks.

The principal constituent of limestone is calcium carbonate and the most stable form of calcium carbonate under normal conditions is calcite. Limestones were usually deposited under marine conditions, so it is surprising that they are not composed of the rarer calcium carbonate mineral, aragonite, which is the normal calcareous deposit from sea water. However, aragonite is unstable under normal conditions and often changes into more stable calcite. Small amounts of aragonite may be found in more recently deposited limestones, but become rare or non-existent in older deposits.

Some limestones contain large amounts of dolomite – calcium magnesium carbon-

Jurassic limestone, Loch Aline, Argyll. The fossils are Gryphea.

ate $(CaMg(CO_3)_2)$ – whilst others, the magnesium limestones, carry significant amounts of magnesite $(MgCO_3)$. Occasionally, appreciable amounts of iron tinge the dolomite limestones a shade of pastel pink, as in the Italian Dolomites and some limestones of Cumbria.

Many limestones are composed almost entirely of fossils. The fact that most of the fossil remains are calcareous by nature does little to alter their composition, although such limestones are said to be of an organic origin in contrast to chemically precipitated varieties. Even so, most fossil remains are not found in an original state, many having been chemically altered after deposition.

Limestones composed of fragments from fossiliferous remains are typically found in the Lower Carboniferous succession, where the Great Scar Limestone of the northern Pennines achieves hundreds of metres of unbroken succession and, as we shall see, provides many renowned examples of limestone scenery. Perhaps the most fossiliferous of all limestones are the Carboniferous reef-knoll limestones which were originally thought to be the remains of coral reefs. However, recent work and analysis of their fossil content, which is often almost exclusively crinoid fragments (with smaller amounts of shell particles), has suggested that knolls are tidal or current-formed lime banks. They provided

Carboniferous knoll limestone. Worsaw Hill, Clitheroe, Lancashire.

a fertile environment for the growth of crinoids.

Most of the Lower Carboniferous limestones are not fruitful strata for the fossil collector who requires whole fossils. Other limestones – the biomicrites – such as the Wenlock Limestone or the Durness limestone may contain whole fossils in a fine grained matrix.

Oolite limestones, usually Jurassic in origin, are of a puzzling round-grained, fish-roe structure. When an oolith is split and viewed under a microscope it proves to be pearl-like in appearance with a series of concentric aragonite layers enveloping a minute particle of sand or shell fragment. Oolitic limestones are still being created in the Bahamas, where sea water saturated with calcium carbonate precipitates layers of calcareous deposits around sand grains. Occasionally, coarse grained ooliths – pisoliths – are found. These accumulate around a larger nucleus by similar processes.

Chalk is the softest and finest of the limestones. Most chalk contains over 90 per cent of calcium carbonate. This is thought to be of organic derivation. Under a microscope, at high magnification, the individual plates of several varieties of algae may be seen. Some chalks contain fragments of other life forms, in particular foraminifera, to such an extent that they were thought to be wholly derived from the skeletal remains of these animals. However, the consensus of opinion has gravitated towards most chalks being formed of algal coccolith ooze which settled on the sea-bed in still water at depths of less than 200 metres. Our best examples of chalk coastal and downland scenery were laid down in Cretaceous times.

An interesting feature of chalk rocks is the occurrence of flints – the farmer's bane in regions of chalk-derived soils – interspersed throughout the chalk in oddly misshapen lumps and nodules. Flint is almost entirely silica, derived from the pointed spicules of numerous sponges which were embedded in the chalk at the time of formation. Later solution of the spicules, formation of gels and redeposition in nodules followed. Flint and related chert are referred to as siliceous rocks. In regions where suitable building stones were unavailable, flint was often the only form of building material. It also provided a useful source of hard material for early man's first attempts to make sharp weapons and tools.

Possibly of equal importance in man's progress, the carbonaceous deposits – coal, lignite and peat – provide us with a readily available source of energy. Most of our coal deposits were formed from the remains of giant plants in Carboniferous swamps. Covered by later sediments, compressed and fossilised, they hardened into the coals of today. More recently deposited beds of fossilised plant remains, the lignites or brown coals, have not yet reached the hard stage of coal (often called humic coal). The peat beds of our moorland areas represent the early stages of coal formation. If allowed to remain and accumulate they could well supply the people of 200 million years hence with high quality coal.

Other non-clastic rocks are of similar utility. The evaporite deposits of gypsum (from the Midlands and northern England) supply us with the raw materials for plaster making. The rock salt beds of Cheshire have been mined and pumped for thousands of years. Both were deposited as precipitates when areas of water dried out to become salt pans, like the Bonneville salt flats of the United States. They were later covered and protected by clastic rocks.

Some iron ores are of sedimentary origin. Certain varieties of bacteria feed on iron compounds, producing brown iron oxides. Over a period of many years, masses accumulated on the beds of lakes and seas forming rich deposits of iron ore. Other sedimentary iron deposits were washed out of upper strata and redeposited as cavity fillings below.

Both igneous and sedimentary rocks may be changed by high temperatures or pressure into metamorphic rocks. As we delve further back in the geological succession there are fewer unmetamorphosed rocks to be found.

Metamorphic Rocks

Each igneous or sedimentary rock has a metamorphic equivalent. Sometimes, when a rock has been greatly altered, it is difficult to ascertain from which original rock it was derived. Composition is an unimportant factor in identification; structure is of far greater importance.

Sedimentary argillaceous rocks may be changed by compression and heat into slates. After metamorphism the original clay particles are no longer recognisable and the rocks must be referred to as pelitic in structure (fine grained) to differentiate them from the psammitic (coarse grained) rocks. Slates are harder than shales and, usually, more fissile. Whereas a shale tends to split parallel with its bedding plane, a slate still cleaves in one direction but this may be at any angle to the original bedding plane.

Cleavage is influenced by the metamorphic pressure. The slates of North Wales exhibit the finest cleavage. Lake District slates are less reliable, but make adequate roof slates and good building stone. The slates of north-west Scotland often exhibit the ripple marks of metamorphic pressure cleavage. They are dark, grey-black slates with an abundance of golden pyrite crystals which corrode with time, leaving rusty holes. The purple, green and blue colours of some Welsh and Cumbrian slates are attributable to small quantities of chlorite or other minerals.

Coarser slates, with the lustrous sheen of indistinguishable mica crystals, are classified as phyllites. In some phyllites it is just possible to see finely differentiated layers of quartz and feldspar crystals. When such layers are more strongly delineated the rock is classified as a schist. A good specimen of mica-schist glitters like a Christmas card, with diaphanous layers of mica crystals defining the cleavage planes. Mica-schists represent the most metamorphosed form of

Slate bridges, Crummack, North Yorkshire.

These Silurian slates are fissile at 90 degrees to their bedding plane, which dips from left to right in the picture.

argillaceous rock. Some Scottish schists have been metamorphosed several times over; each time the direction of cleavage has been changed, producing new alignments of the mica crystals. Some schists are the result of metamorphism of basic igneous rocks. Others have been so strongly metamorphosed that high temperature mineral crystals, such as garnet, are produced. The garnet-rich mica-schists of Perthshire are a good example.

In contrast to the finely banded fissile schists, phyllites and slates, the gneisses exhibit coarse, poorly developed banding of alternate mica and quartz-feldspar layers. Almost like a banded granite in appearance, these coarse rocks are the result of high grade regional metamorphism. Although the mineral composition of gneisses is similar to that of schists, it is impossible to judge, in most cases, whether a gneiss is of sedimentary or igneous origin. Some, indeed, are the product of high temperatures and injection of granitic magma.

Marble is recrystallised limestone, resulting from high temperature (thermal) metamorphism. The original structures, and any fossils, have been completely destroyed and the resultant almost indiscernible micro-crystalline mass produces a hard, durable but workable stone beloved of sculptors. Some so-called marbles – like the dark fossiliferous 'Dent marble' of

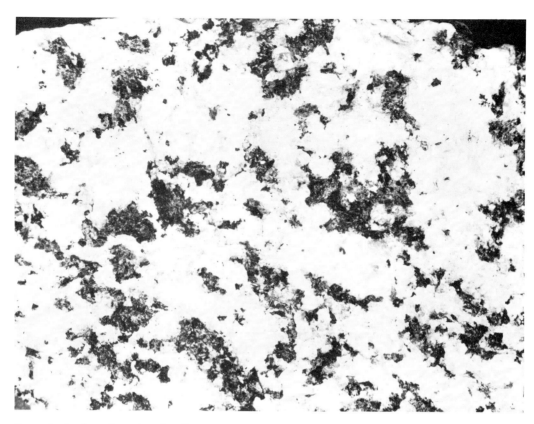

Granodiorite from Strontian, Argyll.

North Yorkshire – are not metamorphic rocks at all. The fact that their fossils are unchanged precludes this possibility.

Metamorphosed arenaceous rocks, containing large amounts of quartz with little else, are known as quartzites. Their crystals may be fused together into a hard interlocking matrix or they may have been completely recrystallised into an almost amorphous and extremely hard mass. Some of the old Cambrian rocks of the Scottish Highlands have been changed into glistening masses of white quartzite.

Whenever a large mass of igneous rock has been injected into the crust, an appreciable area of surrounding rock may be baked by the intense heat of the intrusion.

The amount of heat dissipated into the surrounding rock depends on the mass and temperature of the intrusive body and the thermal properties of the host rock. The surrounding rock will be altered over an area which corresponds to the shape of the intrusion. Such zones of altered rock are classified as metamorphic aureoles. A typical example of a metamorphic aureole, where contact metamorphic zones surround a granite core, is to be found near Skiddaw in Cumbria, where the Skiddaw slates have been baked into a hornfelsed, pale grey rock near the contact with the intrusive Skiddaw granite. Further away from the contact zone the slate increasingly shows a more normal structure until all traces of metamorphism are absent.

It is not only the host rock that becomes altered. The granite itself is finer grained towards the junction and large crystals are absent until, at the junction, it is a very finely crystalline rock. Further away from the junction with the slate, the granite slowly grades to normal. This illustrates the fact that igneous rocks which are cooled more quickly by contact with cooler, surrounding rocks are of finer structure. Metamorphism works both ways, altering both the host and the intrusive rock. Mineral veins, dykes and sills produce similar baking effects, but to a lesser degree because of their smaller size.

Studying Rocks

Rocks are best studied where man-made or natural erosion has produced a large exposed section. By studying large masses of rock it is easier to perceive structures and variations. However, town dwellers who are unable to escape regularly to the country may derive some satisfaction by searching churchyards and identifying the stones used by monumental masons. Large public buildings, shops and banks are often faced with a variety of stones.

Collected pieces of rock may be ground and polished by a similar method to that outlined in Chapter 2. It is always wise to note the location at the time of collection so that the specimen may be more permanently labelled at home.

One word of warning which applies to

Exotic stones like marble are often found in churchyards.

the collection of rocks, minerals and fossils: never dispose of unwanted specimens in a locality other than the one in which the specimen was collected. Careless disposal could lead to many problems for another collector.

Our study of rocks is not complete without an explanation of their stratigraphy and structure. Chapter 4 continues the theme and describes the major structures and their origins.

Coloured Eocene sands, Alum Bay, Isle of Wight.

(Above)   Green malachite, an ideochromatic mineral, is a copper carbonate. The
rusty colour is limonite iron ore.

(Below)   Purple fluorspar cubes.

(Above)   Blue azurite copper ore.

(Below)   Bornite and chalcopyrite copper ores.

(Above)   Heat effects around a mineral vein. The slate, originally grey, has been baked red by heat from the vein.

(Below)   Top: Silurian banded whetstone, Crummack, North Yorkshire.
Bottom: micro folding in Dalradian schist, Argyll.

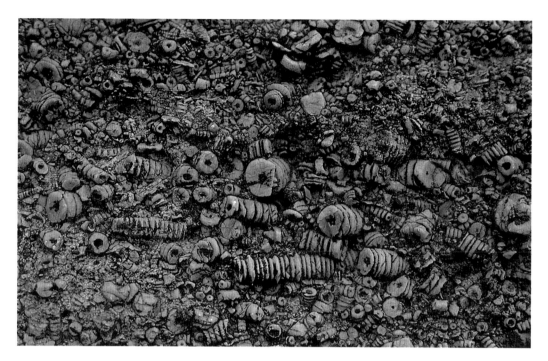

(Above)   Crinoid stem fragments in
Carboniferous knoll limestone.

(Below)   Quartz encrusted ammonite from
Jurassic strata.

(Left)   Steeply dipping limestone,
Twiston, Lancashire.

(Above)   Carboniferous limestone lies
unconformably above Silurian greywacke,
Arcow Old Quarry, Helwith Bridge, North
Yorkshire.

(Above)   Anticline in Carboniferous
limestone.

(Below)   Joints in limestone are easily
eroded by water.

(Above) Erratic boulders, Norber, North Yorkshire.

(Below) Limestone scree, Austwick, North Yorkshire.

(Above)   Anticline in greywacke, Arcow
Quarry, Helwith Bridge, North Yorkshire.

(Below)   Clints and grikes in limestone
follow the course of original joints.

(Above)   Hexagonal columnar structure in basalt, Staffa.

(Below)   Belnahua – a ruined slate island in the Hebrides.

(Above)   Heaps of crushed greywacke for road building.

(Left)   Rusted ladders, the only means of access to this old fluorspar mine.

(Right)   Abandoned slate workings, Tilberthwaite, Cumbria.

(Above) Striding Edge, Cumbria – a glacial arête.

(Left) Blind Tarn, Cumbria – a glacial corrie tarn.

(Right) The Glencoe boundary fault crosses the hillside diagonally above Loch Achtriochtan.

(Above)   Rannoch Moor – a glacially excavated lochan, bog and heather wilderness.

(Below)   Gordale Scar, Malham, North Yorkshire.

4 Structure

Sedimentary Structures

The structure of sedimentary rocks was determined when they were deposited as sediments on the beds of ancient seas and lakes. Many of the calcareous rocks have attained great thicknesses of uninterrupted succession. The Great Scar Limestone of the Lower Carboniferous and the later Cretaceous chalks may reach hundreds of metres of thickness without significant interruption by other rock types. With only minor exceptions, their conditions of deposition remained stable for several million years. Water temperatures were fairly constant. There were no large tectonic events; just gradual slumping and compression as the weight of extra sediments was added from above. Biological conditions and water currents also changed little.

On the other hand, strata of the Upper Carboniferous Coal Measures demonstrate a completely different sedimentation pattern. They exhibit all the classical features caused by cyclical changes in the conditions of deposition. Much of the stratification of the Upper Carboniferous succession was caused by the advance and retreat of the estuaries of the great rivers which flowed from northern regions. A typical Coal Measure succession is:

9   coal
8   seat-earth
7   sandstones (possibly current bedded)
6   mudstones
5   sandy shales
4   coal
3   seat-earth
2   sandstones (cross-bedded)
1   marine mudstones

Firstly, marine mudstones were deposited as a fine silt in water of medium depth far from the river estuary. As sediment built up and uplift of the sea-bed occurred, the river extended its estuary and deposited larger particles of sandy debris over the same area. In some cases these may have been inter-bedded with thin layers of mica flakes, indicating a period of very calm depositional conditions. Eventually the sand-stones built into a delta, rather like the present Danube delta, through which the river had to thread its way to the sea or lake. Swamp conditions prevailed and coal-forming vegetation flourished for a while, rooted in a seat-earth clay from which most minerals were slowly extracted by the vegetation above. Later, the land sank again and a new cycle of marine and estuarine deposits was laid down on top of the fallen vegetation. The whole process was repeated many times during the coal-forming period (Westphalian), giving a fairly predictable repetitive succession.

A similar succession may be found over a wide area of Britain and Europe where the same rhythmical variety of stratification occurs. Thicknesses may reach the order of a few centimetres or a few metres, but there was never any prolonged stable period when a great depth of rock could be laid down. All such rocks began as more or less horizontal strata, although many were folded and tilted later.

Cross-bedding

Occasionally, strata do not conform to the usual parallel stratification pattern. Some bulge into the rocks below, others appear to be bedded at an angle to the surrounding rocks. Sandstones seem to be most affected by this process of 'cross-bedding', because they were often deposited in turbulent water or at shallow depths where strong currents amassed banks and carved gulleys. Cross-bedded sandstones are useless to quarrymen who require large blocks of sandstone with easily worked, parallel bedding planes. Many good examples of cross-bedded sandstones are, therefore, found in quarries because, being useless as a workable stone, they were left *in situ*.

Rather than lying parallel with surrounding strata, cross-bedded rocks appear to be bedded at an angle to the rocks above and below. Delta conditions are an ideal environment for the formation of cross-bedded sandstones. Here, layer upon layer of sediment protrude successively further out to sea, building one sloping sand ramp on top of another. Wind deposited sand dunes also build into cross-bedded structures. This is known as 'Aeolian' or wind bedding.

Ripple Markings

A related structure – ripple markings – appear wave-like in cross section. They are formed as a result of fluctuating currents rippling the surface of deposition. Wave ripples, similar to the structures found on

Cross-bedding in Pliocene shell sands, Suffolk.

Ripple-marked sandstone on a 300 million year old fossilised beach.

A 'load cast' where the weight of sediments has caused
subsidence into the lower beds.

many beaches after the turn of the tide, are often preserved in sandstones. At one time these ripple-marked sandstones were discarded by quarrymen, who required smooth-surfaced flagstones for paving. However, fashions change and ripple-marked flagstones are now in such demand that ripple-surfaced concrete pavings are produced to satisfy the shortfall in demand.

Cleavage

Another feature of some sandstones and shales is their ability to split along a line parallel to their original bedding plane. This phenomenon is attributable either to 'lamination' by pressure from strata above or to thin layers of inter-bedded mica which enhance fissility as a result of their flat, platey structure.

On the other hand, some metamorphic rocks may be fissile at any angle to their pre-metamorphic bedding plane. Their usual cleavage direction is perpendicular to the dynamic forces which caused metamorphism. This imposed cleavage is known as foliation.

Unconformity

The Coal Measures succession, and all other geological successions, are subject to interruption. As we have seen, it is usual (as a natural consequence of gravity) for strata to be deposited horizontally with bedding planes that are approximately parallel. However, some deposition is interspersed with uplift, sinking or folding.

When previously consolidated strata are folded and uplifted above sea level they become, as a result, exposed to the processes of erosion whose general effect is to bulldoze and smooth them into an undulating plane. When such an eroded, contorted

land slumps beneath the sea again – as happened in the case of Britain's Lower Palaeozoic strata – deposition of new sediments takes place on top of the former land area. The new sediments will be deposited horizontally, at an angle which is at variance with the dip of the folded strata below. A junction of strata with dissimilar dips is said to be a plane of unconformity. When uplifted and exposed by later erosion, the unconformity may plainly be seen as a geological time gap in the order of succession.

Dip

The angle between the bedding plane of a stratum and the horizontal is termed the 'dip'. To measure the dip a clinometer is used. In its simplest form a clinometer consists of a flat base which is placed on the bedding plane. Rigidly mounted on the base is a protractor. A pivoted, freely swinging, pointer indicates the angle of dip on the scale of the protractor when the base is placed at the steepest position of dip.

A clinometer is an extra load to carry in the field and, if only the direction of dip is required, rolling a rounded stone or marble down the rock would serve the same purpose and then the angle of dip could be roughly estimated. If more exact recordings are required, a clinometer should be used to give an accurate angle of dip, then a bearing of the direction of maximum dip checked on a compass. Professional apparatus would include both facilities.

Once the dip of a stratum has been found, its 'strike' can be ascertained by taking a horizontal line at right angles to the direction of dip. For instance, in the case of a rock dipping at any angle in a westerly direction the strike would be aligned north to south.

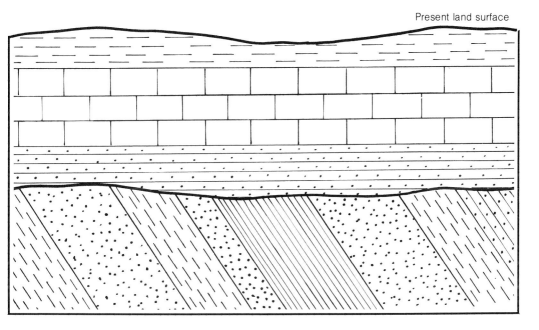

Present land surface

An unconformity – almost horizontal strata lie unconformably above
steeply dipping older strata.

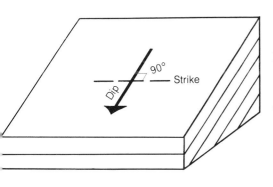

Block diagram of dip and strike.

The effects of dip and strike also influence the way in which an observer views
outcrops of strata in the landscape. A raised
block of horizontally bedded strata – i.e.
with scarcely any dip – would look like a
layered cake or a stack of planks of wood
when viewed from any side. A similar block
of strata dipping away from the observer's
position would have a similar appearance
until viewed from the side when the
maximum angle of dip would be seen. If the
geologist was looking at a hill formed of
strata which dipped towards him at an
extreme angle, the nearest strata would
appear to extend almost over the entire
front slope of the hill. A river flowing
towards the observer would erode through
the hill and expose V-shaped sections of all
the underlying strata.

The solid line shows the dip and the dotted line shows the strike.

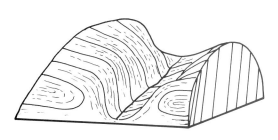

Steeply dipping strata give V-shaped outcrops downstream.

A further problem which may be encountered when strata dip steeply is the determination of whether they have been tilted near to or just beyond the vertical. This knowledge is of invaluable assistance when recording the correct order of succession of steeply dipping strata. An incorrect decision would completely reverse the order of stratigraphy. Fossils, if available, are an important source of evidence in resolving this type of dilemma. However, there are many cases where metamorphism has completely destroyed all traces of fossils, and the only way to solve the problem of succession is to work out how the rocks have been folded and faulted. The succession of some of the older and several times tilted rocks of the Scottish Highlands is still arguable.

Folding

Few sedimentary rocks remain for very long in their horizontally bedded position. Many of the earth's older rocks have been subjected to deformation by forces of four main types.

Firstly, much folding results from deep-seated earth movements in an approximately horizontal plane. These forces are so powerful that the force of gravity is defeated and vast areas of the earth's crust become squeezed, contorted and uplifted. The European Alps were, and are still being, raised slowly but steadily by such processes. Northern Scotland also continues to rise very slowly.

Secondly, as a consequence of the major uplift of some landmasses into mountain ranges, a secondary process of sinking occurs when some strata are unable to maintain their new position. Slumping and slide-folding are the result of such 'gravitational tectonics'.

A third type of folding process may be instigated by the intrusion of deep-seated igneous material or by the injection of a soluble material such as salt. In both instances a dome results.

A fourth and minor cause of folding is produced by the slumping and compaction of partially unconsolidated, newly deposited, sediments.

The simplest fold structures are symmetrical wave forms comprising upfolds and downfolds – anticlines and synclines – arranged in a corrugated pattern. The components of folds are designated the arch limb, trough limb and middle limb (or septum). Simple folding is the result of moderate lateral pressure. Whenever folding occurs, the central part or core of an anticline becomes compressed but the top part of the arch may be subjected to tension and stretched, with consequent cracking and fissuring – forming suitable channels for mineralising solutions. Cavities may be produced beneath the arches of anticlines where the topmost strata are thrust upwards above the lower strata, producing hollows which may also be infilled by mineralising solutions. The saddle reef mineral deposits at Bendigo, in Victoria, Australia, were produced in this way.

Not all folds may be categorised as simple or symmetrical types. In fact, very few folds are entirely symmetrical. Monoclines consist of one limb with strata dipping at a near vertical angle and the other limb with a much less severe angle of dip. Another variation, the recumbent fold, probably began as an ordinary anticline but was then subjected to extra lateral pressure until its arch was pushed over and the two limbs dipped in the same direction. An isoclinal folding pattern is the result of several recumbent folds together. The unfortunate result of recumbent folding is that the upper limb of the fold displays the geological succession in correct sequence. Underneath, on the lower limb, the succession is entirely reversed with the older strata on top. What a problem for the field geologist, particularly if the upper limb of the fold has been eroded away and only the reversed strata of the lower limb remain!

So far we have considered the appearance of anticlines and synclines as seen in cross-section. Viewed from above, the waves and troughs of a recently formed fold system would resemble the frozen surface of the sea. At this point the comparison

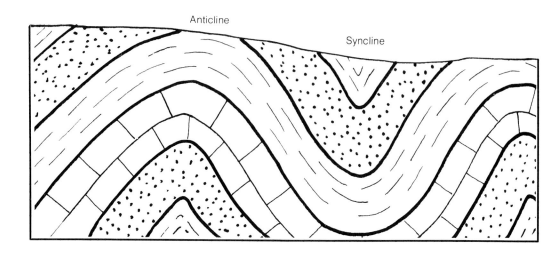

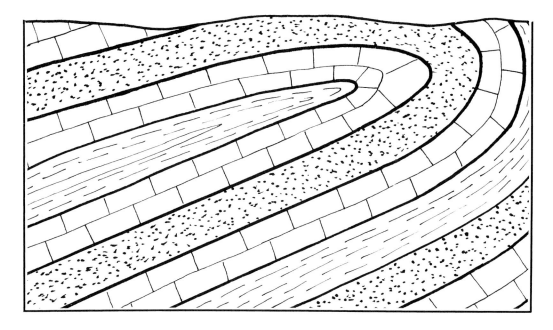

Top: an example of folding. Bottom: a recumbent fold.

ends for, whereas the axis of the crest of a wave would be approximately horizontal, the axis of a fold would tend to pitch downwards. Few anticlines have a horizontal axis. If a north-south orientated anticline, with a limestone core and sandstone crest, had been eroded, the limestone would be exposed at the surface along the crest of the fold. It would, therefore, be possible to trace the anticlinal axis across a tract of land by observing the long, narrow outcrop of exposed limestone. If the anticline pitched downwards in a southerly direction, the limestone band would be seen to narrow as the observer progressed southwards. Eventually, the limestone outcrop would narrow to extinction at the point where the anticline had pitched to such a depth as to bring down the sandstone crest to the present land surface. Further erosion would, of course, reveal more limestone. The angle of pitch of the anticlinal axis is measured in the same way as an angle of dip, i.e. from the horizontal.

An anticline with an axial pitch at either end would appear as an elongated dome. Domes may be formed as a consequence of deep-seated, igneous activity. Granite often has a lower specific gravity than the rock into which it is intruded. Consequently, a large granite pluton surrounded by denser rock is not in a state of dynamic equilibrium. Although held down and enclosed by thousands of metres of overlying strata, a large granite mass would naturally tend to 'float' upwards through the surrounding strata in order to reach a position of isostasy or equilibrium in accordance with its comparative buoyancy. The situation is analogous to what would happen if it were possible to tow a large iceberg down to the floor of the ocean to be tethered and then suddenly released. The iceberg would leap upwards until it reached a position where it could float with the greater part of its mass below the surface.

Granite plutons behave in a similar way except that the process takes millions of years and the overlying strata bend upwards into a dome rather than allowing the pluton to break through. However, erosion may complete the work and reveal the granite core. Granite masses beneath the English Lake District, and the Weardale granite beneath the northern Pennines, both impart isostatic buoyancy to the areas which they underlie – literally holding up the hills above.

Salt

Salt intrusions may also be the cause of dome structures. Salt plugs rising from great depths have forced their way to the surface bringing with them, in some cases, fragments of earlier rocks from great depths. In north-west Germany salt plugs have forced their way through the crests of domes and anticlines.

Cheshire salt is an evaporite deposit from the bed of a dried up sea which later became buried under more deposits. By mining and solvent extraction methods the salt companies produce large cavities beneath the Cheshire plain. Subsidence and slumping produce man-made synclines – the Cheshire salt flashes – which flood to form small meres, an ideal habitat for wildlife. Basins are produced naturally when compaction of loosely bedded material occurs or, in a similar fashion to the Cheshire salt flashes, when ground water dissolves soluble minerals, leaving cavities into which overlying strata collapse. These slumpings are of relative insignificance compared with the results of orogenic movements when major forces heave and fold strata into gargantuan structures.

Earthquakes

When strata are subjected to greater compressional forces than their elasticity can cope with, and the production of folds is insufficient to absorb the strain, the outcome is shearing and breakage along the line of greatest weakness. Sudden movement causes an earthquake and the effects are often devastating. In the San Francisco earthquake of 1906, 700 inhabitants of the city were killed. This figure is horrifying enough when added to the destruction of the city, but is insignificant when compared with the Shansi earthquake of 1556 which resulted in the death of 830,000 Chinese. However, the magnitude of an earthquake cannot be judged by the number of its victims. This, to a large extent, depends on the population of the affected area.

The usual way of measuring the magnitude of the shock is based on the seismographic measurement of the maximum amplitude of surface waves over a period of about twenty seconds as units on the Richter scale. Each additional whole number on the Richter scale denotes a tenfold increase in earthquake energy. There are other systems. The Rossi-Forel intensity scale ranges from 1 to 10. On this scale an earthquake of intensity 1 would be imperceptible to seismographic apparatus. To qualify for a grade of 10 an earthquake would have to cause 'great catastrophe'. The Mercalli earthquake intensity scale ranges from 'not felt except by a few people in favourable circumstances' to 'panic' and 'total destruction' at the maximum intensity of 12.

By using seismographic equipment it is possible to predict an earthquake. Usually there are preliminary stirrings and rumblings deep down. Then, after each major event, there is a settling period of progressively weaker after-shocks. In earthquakes of low magnitude there is little apparent damage. Only in major earthquakes are pipelines broken, buildings destroyed and faults displaced by several metres.

Faults

We live precariously, building our homes and settlements on huge floating rafts of rock; fragmented pieces of crazy paving floating on a molten sea. Some 300 million years ago the land which became Britain was situated near to the equator. Since those times our particular raft, or plate, has travelled thousands of miles north of its original position. In the process of moving there has been much jarring, bumping and deformation as the gigantic plates jostled for position.

Just as in a storm-ravaged harbour packed with boats, there have been casualties. Some plates were compressed and pushed under others, to be re-absorbed and partially melted into new volcanic material. Others were smashed into smaller fragments, then welded together again by the same forces. Some were more pliable and suffered little fragmentation even though distorted and folded.

Forces continue to change and re-form the earth's crustal plates. There are three types of plate: continental plates, oceanic plates and those which cover a mixture of the two environments. Constructive and destructive plate margins have been described in the last chapter. There is one other type of plate margin – the conservative margin, where the two adjoining plates slide past each other in a sideways fashion, neither losing nor gaining material although causing tremors and earthquakes whenever

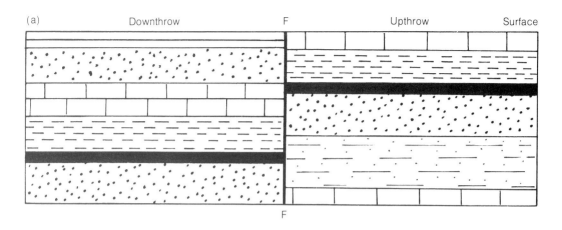

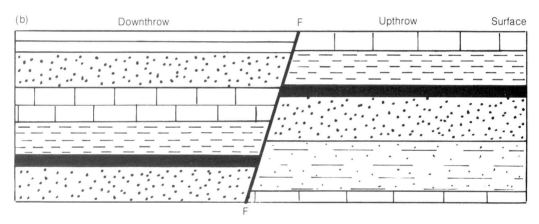

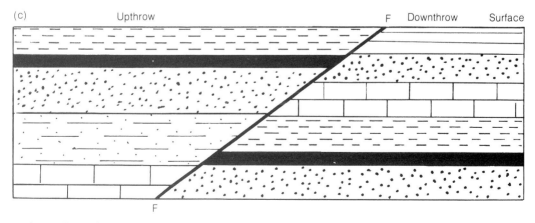

Faulting of a coal seam: (a) normal, vertical fault plane; (b) normal, inclined fault plane; (c) reversed fault.

movement occurs. The San Andreas Fault of California is a notorious example of this type of fault.

To a limited extent earth movements are taking place slowly and unnoticeably all the time. On the other hand, rocks have a certain elasticity and plasticity and do not respond to forces until such time as they are unable to absorb any more compression and folding. Then, with sudden violence, they buckle until a fault fissure yields and begins to move. In a matter of hours or days the opposing sides of a fault like the San Andreas may undergo considerable lateral displacement until all the tensions are relieved. A quiet period then ensues, until breaking point is reached again.

Within the boundaries of the plates there are smaller faults. Many have not been displaced for millions of years. They may never move again, but there is always a possibility that an old fault, as a line of weakness, will once again respond to external forces. Some faults are insignificantly small, having a 'throw' of less than a metre. Others may have displacements of hundreds or thousands of metres. Faults are classified in the following categories, which were an attempt by the mining profession to rationalise a large number of imprecise terms into a universal terminology. Miners tend to be paranoid about faults which may, when encountered, bring about the premature termination of a profitable coal seam or mineral vein. Having lived all my life in a mining town I have seen the closure of several coal mines due to difficulties arising from faulted seams.

Normal fault

Normal faults are the most abundant type of fault. They are the result of tension or low horizontal stress, plus great vertical

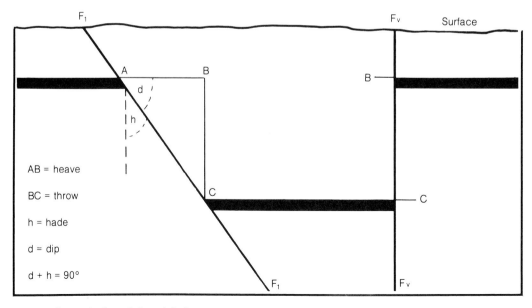

AB = heave

BC = throw

h = hade

d = dip

d + h = 90°

Heave, throw, hade and dip of faults.

stress. The outcome is that one block slides down the plane of faulting and becomes displaced to a lower position than it occupied before faulting. Conversely, its opposing block seems to have been displaced upwards. Geologists refer to the latter as the upthrow side of the fault and to its neighbour as the downthrow side. The amount of vertical displacement is the throw. In mine workings a faulted vein or seam would be said to be upthrown or downthrown by an amount measured at right angles from the point where the vein meets the fault to its new position on the opposite side of the fault.

Unless the faulting is vertical, there must also be some horizontal displacement when one block slides down the inclined fault plane. The amount of horizontal displacement is termed the heave. This varies according to the amount of throw and the angle between the fault plane and the vertical – commonly known as the hade. An alternative way of measuring the angle of a fault would be to measure its angle of dip between the fault plane and horizontal.

Reverse fault. Corresponding strata are marked X. Note the calcite mineralisation at the top left.

Reversed fault

When earth movements produce high horizontal pressure and low vertical pressure, the opposite of a normal fault – a reversed fault – results. Many reverse faults have a high angle of hade because of the high horizontal formative stresses, and may be nearer to the horizontal than the vertical. The upthrow of faults of this nature may be thrust over the downthrow side for a considerable distance. Such a fault is called an overthrust. A good example – the Moine thrust – is situated in the north-west of Scotland.

Wrench Fault

A third type of fault, the wrench, tear or strike-slip fault, has little or no vertical displacement, but a large degree of horizontal movement along the strike of the fault. The Californian San Andreas Fault again provides an excellent example of a still active fault. From a simplistic point of view, if a house was built straddling the fault it would not be many years before the householders had two properties for the price of one.

Nearer to home we have the Great Glen Fault of the Scottish Highlands. Two types of granite, the Strontian granite and the Foyers granite on the north and south sides of the fault respectively, have been proved by analysis to be the same rock. They are now 65 miles apart! This phenomenally distant separation was not achieved in a matter of hours. Some faults remain active for hundreds of millions of years. When movement occurs it is quite rapid, and the fault may be displaced several metres in a few hours. For many years nothing more happens until sufficient stresses have been built up to move the fault once again. Even faults with no recent history of movement may become re-activated. Others have never stopped moving and have progressed at an imperceptibly slow rate for many years.

Multiple Faults

So far we have considered faults as having one plane of fissuring or breakage. For study purposes it is advisable to consider the plane of faulting as being a single entity. In most cases, however, the shear zone is not so well defined and there are often confusing minor faults running in echelon with the strike of the main fault. A glance at the heaved and cratered battlefield appearance of the land surrounding a major fault zone is enough to confirm this.

A fault is not a clean break, but a torn wound between its two opposing sides. The heat of friction and movement may have caused metamorphism in the immediate vicinity. Crushed and broken pieces of rock-fault breccia are found within the fault fissure and its walls may well be slickensided by friction. Slickensides are linear abrasions caused by the rubbing together of the two fault wallrocks. Their streak pattern indicates the direction of movement of the fault.

When several faults are formed in parallel, so that each results in a progressive increment of downthrow or upthrow, the result is step faulting. The Craven faults of Yorkshire have three branches – the North, Mid and South Craven faults. Each one raises the Great Scar limestone progressively higher towards the north. If normal faults dissect the landscape, as in the Midland Valley of Scotland, a system of block faulting is produced.

Tensional forces pulling at the earth's crust produce an effect known as a 'graben' where a large V-shaped wedge of crust sinks into the rift formed by tension. The Rhine follows a graben for a great deal of its course. The zones of faulting are situated laterally along the edges of the Rhine valley.

Joints

The term 'joint' is derived from the jargon of coal miners, who likened the cracks and fissures they found at right angles to the bedding planes of coal-bearing strata to the joints between bricks in a wall. To avoid confusion with faulting, it must be remembered that joints, unlike faults, have undergone no displacement in any direction parallel to the plane of fissuring. Igneous rocks may cause confusion in this respect, as there may be no visible evidence of displacement even if it has occurred.

Although joints and faults are regarded as separate entities, the former may be a product of the tectonic forces which caused the latter. Joints may also be considered to be products of folding or displacements by intrusions. Some joints are found only near the surface and may be attributable to freezing or water erosion. More often than not, however, these two processes are instrumental in widening joints which are already present.

Jointing may conform to the same pat-

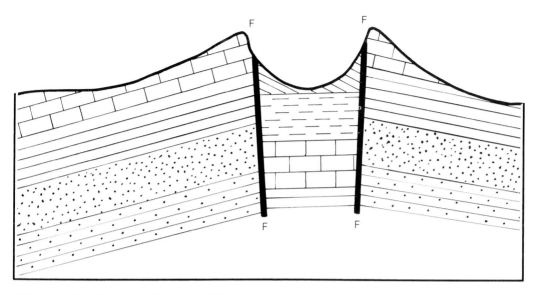

Graben or rift valley in which a central block is faulted downwards.

tern over a wide area. The large pavements of limestone exposures usually have their blocks aligned in a fixed grid system, with their water-widened joints forming a distinctive mesh of open channels. When joints conform to a fixed pattern over a region they are known as systematic joints. If there are two directions of systematic jointing in a region these are referred to as complementary joint sets. In analysing the joint structures of an area it is important to find the joints which penetrate deepest, often passing through several strata, rather than shallow and random surface joints. When the directions of these 'master' joints have been plotted it is often possible to relate their promulgation to other tectonic phenomena by studying geometrical relationships.

Although the majority of joints are of tectonic origin, some are the result of shrinkage due to dehydration or cooling. The columnar jointing of basalts and other lavas can be attributed to shrinkage into hexagonal structures during the cooling

Gigantic blocks of limestone, undercut by a stream, have broken along joint planes and fallen into the stream below.

process. Jointing in coal seams has never satisfactorily been explained. Miners find that hard, black coals split best in one direction to give shiny, glossy surfaces. When broken at right angles to the cleat (best cleavage) a duller surface is obtained. Just like wood, the smoothest surfaces are obtained by splitting with the grain, although it is easier to saw across the grain. Coal is more easily cut by working across the cleat. Another interesting aspect of this form of jointing is that it is remarkably uniform over a large area of coalfield and, indeed, over the entire northern hemisphere where the general trend of cleat is north-west to south-east.

Joints have played an important part in the formation of certain types of limestone scenery, an aspect that will be explained in more detail in Chapter 5.

5 Geology and Scenery

Scenery is often considered to be the only permanent feature in a world of constant change. Gavin Maxwell was so impressed by the ruggedness of the Scottish landscape that he chose *The Rocks Remain* as the title of one of his books. Such apparent permanence is only an illusion and our landscape is constantly undergoing change by the agents of erosion and, occasionally, at a more rapid rate, by folding, faulting and vulcanism. At the moment, Britain is scarcely affected by the latter forces. It is undergoing a relatively quiet period during which landscaping by erosion takes precedence.

The landscape is a product of many contributory factors. Perhaps the most important feature is the nature of the underlying strata. Their hardness or softness determine the degree to which they may be planed down, shaped and smoothed. Grits and granites tend to be durable and resistant. Areas where these rocks outcrop often stand stark and formidable above the surrounding countryside. Soluble, crumbling, or loosely jointed rocks are readily eroded and weathered and therefore tend to form gently undulating landscapes. Folding and faulting are also important factors in the shaping of the landscape.

Although a newly formed anticline starts as a ridge, after millions of years of erosion it is the synclines that usually remain as hills and mountains. This occurs because the rock in the trough of a syncline has been compressed and hardened whereas the crest of an anticline was subjected to

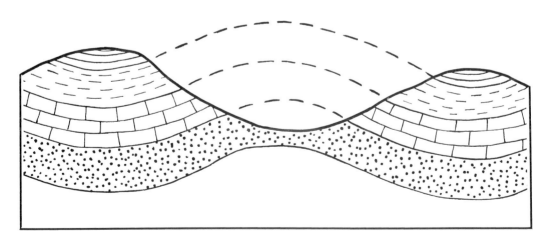

The stretched crests of anticlines are often eroded to valleys, whilst the nearby synclines remain as hills.

stretching and fissuring which imparted structural weaknesses. Such lines of weakness were more easily worn down by rivers and ice. The ice sheets of the final phase of glaciation retreated just over ten thousand years ago – a short time, geologically. Much of Britain's glacial scenery remains in pristine condition, hardly affected by the last ten thousand years of erosion.

Ice Erosion

Although areas to the south of Yorkshire and North Wales were not affected by the last phase of the Ice Age, evidence of earlier glaciation is still to be found in these southern regions as far south as the Severn and Thames estuaries. Ice sheets reached their southernmost limit in Pleistocene times and many of the glacial deposits north of the Severn-Thames line, in the Midlands and East Anglia, date back to these times. 'Drift' deposits are a valuable indicator of the flow direction of the ice sheets. By careful checking and recording of the rock types to be found in glacial drift it is possible to determine where a glacier originated and its direction of travel. Boulder clay, a mixture of finely ground rock flow and pebbles (some of massive size) is the most widespread drift deposit. In some areas it attains depths of many metres.

Glaciers began on the northern hills, usually in the bed of a pre-glacial stream, and eroded their way inexorably downwards, powered by their weight and the force of gravity. Accumulations of snow became compressed and consolidated by the weight of extra snowfall on top. Eventually, as sufficient thickness and mass was added, the pressure from above caused melting of the ice below so that a thin layer of water acted as a lubricant between ice

and rock and assisted faster downward momentum of the ice. Although the ice was fairly bendable and plastic, some cracking and crevassing occurred. As the ice travelled, it picked up stones and boulders which were transported and scraped against the underlying rock surface. The result was plucking, scouring and smoothing as the glacier progressed, gathering ever more debris yet still retaining the previously ground down pieces of rock to be deposited later as boulder clay.

Some debris was dumped along the edge of the glacier as lateral moraine. Medial moraine was deposited in the middle of the glacier and terminal moraine at the end, where melting occurred. As the climate improved and glaciers retreated and eventually melted, the drift deposits were left as markers of their progress. These are thickest where several glaciers joined together to form ice sheets. A study of moraines and striation (gougings) on the rocks which glaciers traversed is another useful indicator of their progression. Erratics – ice transported boulders – were often deposited great distances from source.

As well as leaving moraines and rounded heaps of boulder clay, called drumlins, the glaciers made their mark on the contours of the landscape. Those which descended from a very steep gradient often deepened and hollowed out the land at the bottom of their almost vertical back-slopes, into corries (in Scotland), tarns (in the Lake District) or cwms (in Wales). Universally known as cirques, these upland basins are now often filled with water and drained by an overflow stream at their lip. Slates, schists and granites were eroded into the best examples, of which Bow Fell Tarn in the Lake District is a splendid specimen. When two cirque glaciers began in a back to back formation, each progressing in

Drumlins – hummocks of boulder clay – deposited by ice. Ribblesdale, North Yorkshire.

opposite directions from the crest of a hill or ridge, the resultant backcutting produced a ridge resembling the shape of a hollow ground blade. This feature is called an arête. Striding Edge in Cumbria is a typical example.

Glacial erosion down a river valley which was originally V-shaped produced a U-shaped profile with the original spurs truncated. The hardness of the surrounding rock may have inhibited widening of some valleys, but it is possible that even these valleys could be considerably deepened.

After melting of the ice and a consequent rise in sea levels, many of the glacially deepened river valleys of western Scotland were flooded by the encroaching sea. They remain as fjords or sea lochs. Others, like the inland valley of the River Tummel in Perthshire, were deepened to such an extent that, after the Ice Age, they became flooded. The Tummel valley ribbon-lake is viewed best from Queen's View at the eastern end of Loch Tummel. This wild panorama was one of Queen Victoria's favourite Highland scenes.

The glaciated U-shape of Glencoe.

Loch Tummel, Perthshire. The glaciated valley is flooded by the River
Tummel.

From this viewpoint the Road to the Isles leads the eye through the glacial landscape to Loch Rannoch and, a short distance to the west, the wilderness of Rannoch Moor which belies the fact that granite may usually be regarded as a hard rock. Here, the forces of glaciation have scoured the granite down to a flat, heather peat bog and lochan wasteland surrounded by hills of schists and quartzite. In this instance the hard granite was softer than its two companions so it was preferentially eroded by the ice which took the line of least resistance.

As the Ice Age ended and the ice melted, large lakes of glacial meltwater were often imprisoned between enclosing hills and trapped by dams of unmelted ice. Eventually, the water found a way through and escaped as a raging torrent, eroding a deep gorge in its progress. Glacial meltwater channels often defy the usual rules of nature. Some can be seen running along the crests of hills indicating, perhaps, that the hills were hemmed in by ice on both sides. Others have widened what may once have been the courses of minor streams. One of the most spectacular overflow channels, on the Lancashire-Yorkshire border, is now occupied by the Yorkshire Calder and Lancashire Calder rivers which, along with the railway and road routes, exploit this narrow, naturally excavated east-west route.

Wastwater, England's deepest lake, was gouged out by glaciers.

Canyon eroded by glacial meltwater. Calder Valley at Cornholme, West Yorkshire.

In some instances, where glaciers have deepened river valleys, the courses of tributaries have remained unaffected. After retreat of the ice, the tributaries were left 'hanging' above the main valley and often plunge as magnificent cascades to the valley floor.

Other streams have been diverted by moraine dams and have to reach their confluence by a more precipitous route. Cotter Force in Wensleydale is an example of a waterfall produced by a moraine diverted stream.

Not all ice erosion ended with the Ice Age. Freeze-thaw weathering is still an important agent of erosion even though there are no permanent glaciers in the British Isles. Rocks in wet, cold areas of Britain are particularly susceptible to this type of attack. The steep mountainsides of north-west Scotland, the Lake District and some northern limestone areas provide the best examples of freeze-thaw weathering. South-facing slopes may be the most affected. Water penetrates cracks, fissures and joints in the rock, then freezes during the long, cold winter nights, expanding as it freezes and prising the rocks apart. The next day the sun melts the ice back to water which penetrates a little further down the slightly widened fissures, freezes and expands again the following night. Eventually, the rocks fragment and chunks fall off the high cliffs to collect below as a steeply inclined scree slope. In southern Britain most scree slopes have become grassed over, but in cooler, more northerly regions some scree slopes, such as those below Illgill Head at Wastwater in Cumbria, have probably remained in active scree production since the end of the Ice Age.

The famous screes at Wasdale, Cumbria.

Water Erosion

The Sea

The sea is a powerful agent of erosion, and several methods of attack are included in its armoury. One method is chemical erosion, or corrosion, at contact zones between sea and land. More soluble rocks such as limestone and chalk are readily attacked, particularly where their joint systems allow ingress of water. As the sea water penetrates joints at each successive high tide, the fissures are progressively widened into natural arches and caves. These structures eventually collapse and stacks, separated from the main cliff, remain where once headlands were found.

Stacks are also produced by the physical processes of erosion – hydraulic and abrasive weathering. The sea is constantly in motion, using the energy of wind and tide to batter cliffs, beaches and dunes. As each pounding wave dashes against a cliff, considerable hydraulic pressure is produced which forces water through rock fissures, literally pushing the rocks apart and, at the same time, using sand, pebbles and boulders as battering rams to widen cracks further and erode by abrasion. These weathering processes, coupled with some chemical erosion, rapidly produce caves and undercut areas of cliff. Eventually, the cliffs become undercut to such an extent that collapse is inevitable. The cliffs that are undercut to the greatest extent are those with a small angle of dip, especially if softer strata are situated at their base. Near-vertical cliffs dipping slightly towards the shore offer most resistance because their bedding planes are broadside on to the waves.

None of the removed material is wasted. Most is redeposited further along the coast

Stacks at Bedruthen, Cornwall.

as a result of the processes of longshore drifting, whereby wind-blown waves hitting a beach at an angle transport pebbles up the beach in the same direction. As the waves wash back down the beach, the pebbles roll back perpendicularly to the shoreline only to be pushed further along the beach by the next wave. Pebbles may be carried for considerable distances around the coastline, from north to south on the east coast, south to north on the west coast, and west to east on the south coast.

On Land

Streams and rivers are the water eroders of the land. The amount of water available depends very much on the amount of rainfall and the run-off rate which is, in turn, dependent on the soil texture, ground cover and underlying strata. Streams in mountainous western districts, where there is high rainfall, have most energy to transport rocks and pebbles to scour and abrade their beds.

The load of abrasive particles carried by a stream is greater near its source when the stream flows at a steeper angle. There is a maximum possible load after which the corrosion of new particles is balanced by deposition of old particles when the stream becomes saturated with transported material. A stream in flood may carry hundreds of times its normal load of debris although its speed is only three or four times normal.

However, the rate of erosion also

depends on the hardness of the rocks and, to a certain extent, on the abrasive qualities of their components. A rock composed of a high percentage of quartz in a hard matrix would be more resistant than a rock of similar composition in a soft matrix. For this reason, certain types of sandstone and grit with a soft limy matrix are more easily eroded than many granites, because their hard quartz grains are readily separated and made available for transportation. Conversely, some granites tend to fall apart when their feldspars break down.

As streams progress towards the sea and the profile of their bed becomes less steep, much of their transported debris may be

Gordale Scar, Malham – a product of water erosion.

dropped as alluvial deposits. This produces fertile soils, rich in minerals, derived from upland areas where only leached, infertile soils remain.

Up in the hills there is a tendency for streams to cut backwards into their watershed. Eventually, this cutting back of streams on both sides of the watershed ridge imparts a zigzag appearance to its crest. Constant cutting back slowly lowers the height of a ridge.

The drainage of an upland area is complex. It is partially determined by the original structure of the land. Streams that evolved as drainage channels down the slope of a large upfold or continental arch took the easiest, most obvious route down the dip of strata. Such 'consequent' streams are well illustrated by the rivers to the east of the northern Pennines, which all flow with parallel courses towards the North Sea.

Other streams do not conform to the general direction of dip and may join the consequent (or dip) streams approximately at a right angle. Such streams are the result of denudation by consequent streams cutting down to softer strata. This provides a weak line of least resistance for backcutting and corrosion by small tributaries of the main consequent streams. These streams follow the direction of strike of the strata and are therefore known as strike streams or 'subsequent' streams.

Both dip streams and strike streams flow in a direction dictated by the geological structure of the area which they drain. Some areas, however, have been subjected to uplift and folding since the development of their drainage systems. Where uplift took place at a slow rate, the existing rivers were able to maintain their courses, often being given extra impetus to erode by the raising of their source. The River Indus

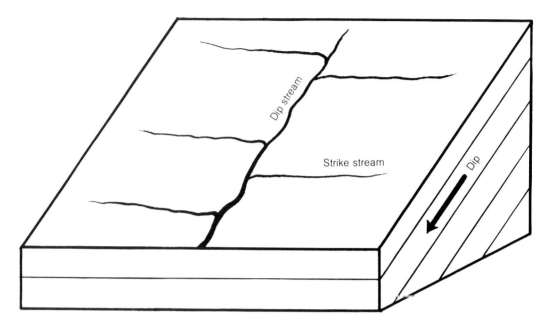

Block diagram – dip streams and strike streams.

flowed from a source situated to the north of the present Himalayas even before the mountains were raised. As the mountain chain slowly built up, the Indus maintained its old course by eroding deep gorges through the mountains so that now it flows through the highest mountains in the world. The European Alps are crossed twice by the Danube which, like the Indus, deepened its bed at a rate fast enough to counteract the slow uplift of the mountains. River systems dating back to a time before the present geological structure of the landscape are said to be 'antecedent'.

Other drainage patterns appear to have nothing whatever to do with the structures over which they flow. In these cases they have succeeded in eroding away all the structure which influenced their early progress down to the level of an ancient structure which had no influence on their directional development. Such a drainage pattern is said to be 'superimposed' on the landscape.

Wind Erosion

The driving force and potency of the wind as a transporter of erosive particles has, perhaps, been underestimated in the British Isles. In the desert regions of the world, high speed winds are capable of sand-blasting and smoothing rocky protrusions into weird shapes. An outstanding example of this type of erosion is the smoothed and detail-less appearance of the sphinx at Ghizeh. Nearer to home there are wind-eroded natural sandstone sculptures on the hills above Nidderdale (Yorkshire), especially at the much visited Brimham Rocks. Joint planes and softer areas of the sand-

Wind eroded grits above Nidderdale, Yorkshire.

stones have been eroded by a mixture of rain and wind-blown sand; the exposed moorland position being a major contributory factor to the sculpting. Some tors on Dartmoor have been similarly eroded.

Wind, with the addition of frost and rain, is also responsible for the denudation of some of our mountain tops. Exposed basalt mountains of Morvern, Scotland, are degradable into fertile soils which support a mixture of hardy grasses and plants. The chances of these colonies of plants maintaining their hold on the friable basalt soils is lessened by dry conditions and high winds that remove the basalt particles to lower lying areas. On the other hand, the basalts are only broken down by severe winter conditions, so the ecology of such exposed areas is finely balanced.

Limestone Erosion

Sedimentary rocks are by far the most abundant in the landscape. Limestone, because of its high solubility and fissured structure, is best regarded as a separate entity and dealt with first. Limestone is most vulnerable to the effects of water along joint planes, which are usually so pronounced that limestone is depicted on geological maps as a brick wall formation.

Limestone is more soluble in weak acids than any other type of rock. Its open-jointed structure only aggravates the problem and speeds the processes of chemical weathering. Weak acids originate when atmospheric carbon dioxide becomes trapped in the pore spaces of the soil. This dissolves in rain water, producing dilute carbonic acid. Further acids are added if peat or humus overlies the limestone. As

91

rain water soaks through peat, or other undecayed vegetable matter, dilute humic acid is produced which, in combination with the dilute carbonic acid from the atmosphere, eats away at the limestone and widens the joints. When the blanketing soils are removed by ice or other types of erosion a limestone pavement is revealed consisting of blocks of limestone (clints) and enlarged joints as gaps (grikes) between the clints.

Pavements are found in most limestone landscapes, but the best developed examples are situated in the Carboniferous limestone districts of North Yorkshire and Cumbria where high rainfall has assisted their erosion. The clints and grikes above Malham Cove and at Chapel-le-Dale, near Ingleton, are justifiably famous examples of horizontally bedded limestone pavements. Further to the west at Clouds, near Kirby Stephen in Cumbria, sloping limestone pavements have been eroded in dipping Carboniferous limestone. The sheltered grikes provide a cool, damp environment for shade-loving plants like lily of the valley and hart's tongue fern. On top of the clints the arid, exposed micro-climate favours sun-loving, hardy species, such as rock rose and a variety of alpine plants.

The solubility of limestone, and consequent widening of its structural weak-

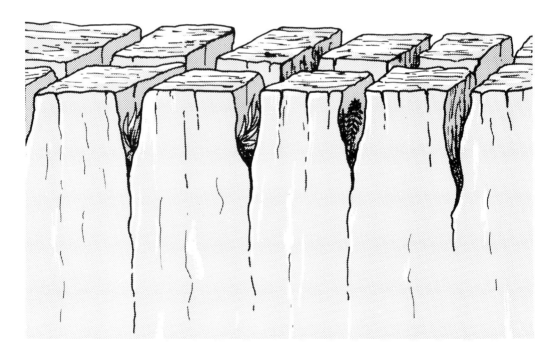

Clints and grikes – water widened joints in limestone.

nesses, is an important factor in the production of its unique formations and weird drainage patterns. Surface water is a rarity in a limestone district. Viewed from above, some areas would appear to be almost desert-like although situated in districts of high rainfall. Streams originating on the gritstone cap rock above the limestones of the Yorkshire Dales flow for several miles then suddenly disappear underground at a 'water sink' in the limestone.

Water sinks are a product of the same chemical process that makes pavements. If a limestone district is overlain by boulder clay deposits, a stream may pass safely across even the most fissured surface, provided all the fissures are blocked by clay. However, as soon as bedrock is reached, the old water course quickly becomes redundant when percolating stream water finds a new route through the joint fissures of the limestone. Eventually the stream will cut back through the boulder clay at the surface, producing new water sinks ever nearer to its source. The end product is a 'dry-valley' in which water flows only in times of extreme flood when underground drainage systems become blocked or saturated.

Examples of water sinks are found at the appropriately named Water End in Hert-

A stream disappears at the Water Sinks, Malham.

Dry Valley at Malham.

fordshire and at the Water Sinks near Malham Tarn in North Yorkshire. The latter stream has noticeably retreated towards its source at the outflow of Malham Tarn in recent years. At one time this stream continued down its valley (now known as the Dry Valley) and plunged over the top of Malham Cove in what must have been a cascade of greater dimensions than Niagara. However, this sight has not been seen in living memory although there are records of a splendid occurrence after a period of extremely high rainfall in the last century.

Disappearing streams cannot remain underground for ever. Their vertical descent is limited to the thickness of the limestone strata, until a junction is reached with an impermeable boundary such as slate. Until that point is reached, vast cave systems may be produced by the underground river; caves such as those in the Mendips, Peak District and Dales that are meccas for pot-holers. Many cave systems are several miles long with amazing convolutions. Some systems cross each other at different levels, never meeting; others are connected by tortuous passages.

There are several theories about the origin of subterranean cave systems in limestone. The most probable explanation of their erosion is, again, the theory of acidity of percolating ground water which may contain several hundred parts per million of dissolved carbon dioxide. Limestone (which is almost pure calcium carbonate) reacts with the acid according to the formula: $CaCO_3 + H_2O + CO_2 = H_2Ca(CO_3)_2$. The resultant calcium bicarbonate is more soluble than the carbonate in water. The reaction takes place more quickly in cool conditions when the limestone is thoroughly saturated.

Pot-holers have observed that some cave systems appear to have been formed at several (usually three) different horizons. This may be explained by the fact that the three major advances of ice during the Ice Age produced three drastic alterations of water table as the land was planed down. The changing water tables resulted in three different levels where chemical erosion could have maximum effect.

What happens to the dissolved calcium bicarbonate? Limestone and chalk districts are notorious for their hard water which furs up pipes and scales the insides of kettles. Here is one clue to the production of lime deposits. Why should the dissolved calcium bicarbonate be re-precipitated as calcium carbonate? There are several reasons. As springs bubble to the surface there is a release of pressure. Pressurised solutions may hold more dissolved bicarbonate than a solution at normal pressure; so precipitation occurs. Also, when a solution becomes oversaturated due to evaporation of some of the water, a pro rata amount of solute must be precipitated. Another possibility is that the energy and oxygenation produced by a waterfall helps to precipitate the dissolved calcium carbonate. In all three cases the calcium bicarbonate breaks down again into calcium carbonate, water and carbon dioxide – the previous formula in reverse. This production of carbon dioxide may account for some caves in limestone areas of France which are filled with carbon dioxide gas.

The products of precipitation are stalac-

Stalactites in the Wolverine Cave, Stump Cross Caverns, North Yorkshire.

tites and stalagmites – calcium carbonate icicles and stumps – hanging from cave roofs and building up from the floors; formed as water droplets evaporate and deposit successive thin layers of calcium carbonate. In a similar way curtains of calcareous tufa grow on the rocks around waterfalls in limestone districts. Petrifying wells exploit this phenomenon; any articles left in the well are not really turned to stone but merely become coated with a deposit of calcium carbonate. Mother Shipton's well at Knaresborough, North Yorkshire, is a famous example.

Springs appear wherever a junction between permeable and impermeable strata occurs. Typically, a junction between limestone or sandstone and slates, shales or

Springs surface where impervious slate outcrops at the base of limestone near Austwick, North Yorkshire.

granites would produce a well-delineated spring line. Springs would be seen along the edge of a dome or anticline at a junction of dissimilar strata or at the plane of an unconformity. In the past the geology of spring formation greatly influenced the development and growth of villages, which were always situated near a good water supply.

Other Sedimentary Features

Just as the jointing and bedding properties of limestone affects its erosion and landscape potential, the similar structure of all sedimentary rocks has a great influence over the type of landscapes they underlie. Horizontally bedded sedimentary strata, first uplifted, then dissected by glaciers and stream erosion, produce a valley and plateau landscape. If the hill tops are overlain by a hard, protective layer of gritstone, as is the case with many of the Pennine hills, the effect is most pronounced. Were it not for

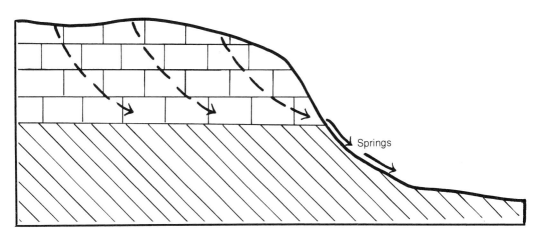

Spring line at the junction of permeable and impermeable strata. Many villages developed near spring sources.

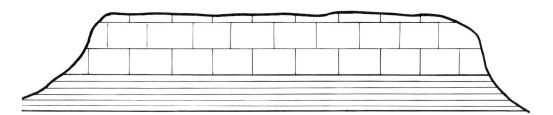

Plateau of horizontal strata.

its exposed position, the summit of Ingleborough in north-west England would make an ideal site for a superb cricket ground, so flat and level is its surface. During the Roman occupation the summit was used as a hill fort by local tribes because of its strategic dominance over the Brigantian landscape. Benn Eighe in Wester Ross similarly stands sentinel above an eroded landscape, its horizontally bedded Torridon sandstones protected by a durable cap of quartzite.

Slightly tilted strata produce a sloping plateau, whereas a greater angle of dip gives a scarp and dip slope formation. The steeply eroded cross-section of the strata is known as the scarp slope and its more gently dipping complementary slope is the dip slope. Strata of variable hardness arranged horizontally or at a small angle of dip may be eroded by water and ice into a series of escarpments progressing up the sides of a valley in a stepped formation. The Yoredale series produce fine examples of

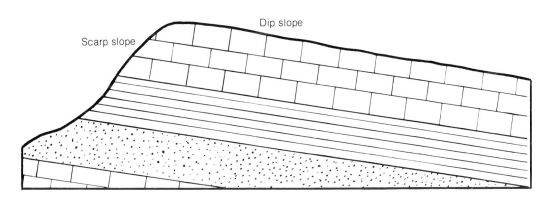

Dip slope and scarp slope.

97

Scarp slope and dip slope. Limestone scenery near Austwick, North Yorkshire.

this terraced structure, and a favoured site for waterfalls is at the edge of each escarpment. Wensleydale and Swaledale are noted for these features, of which Hardraw Force is probably the best known example. Larger examples of escarpments may be seen in the Cleveland Hills and at Lincoln, where the Minster stands on a foundation of limestone surmounting the softer, Upper Lias. Steeply dipping or vertical strata, harder than the surrounding landscape, protrude as ridges or hog's backs.

Faulting of sedimentary strata may throw up hard strata above softer more easily eroded strata. The southern edge of the Askrigg Block, where fault scarps of Great Scar Limestone stand high above the eroded downthrown shales of Yorkshire Bowland, is a natural delineation of the edge of an area of plateau isolated by the Craven fault system. In the Ochil Hills, north of Stirling, resistant andesite lavas stand above the downthrown Carboniferous strata to the south of the Ochil Fault. What a scarp this would have been if no erosion had taken place! The south side is downthrown by 3,000 metres. Even now the scarp is a prominent feature in the landscape, stretching for a distance of twenty miles.

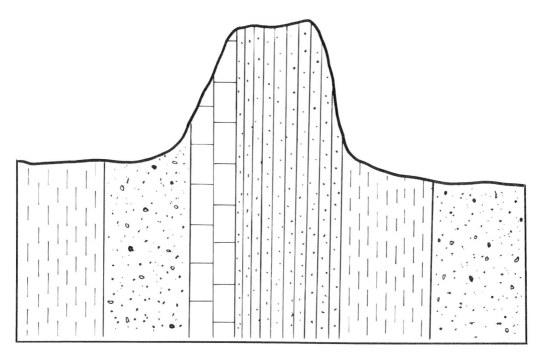

Resistant vertical strata stand above the landscape as a hog's back.

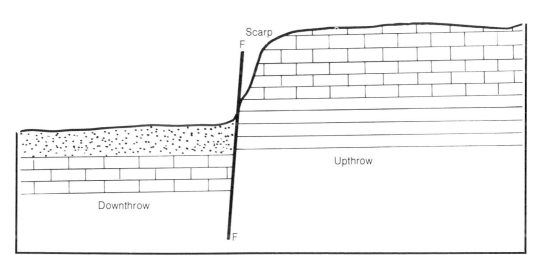

Fault scarp formation in cross-section.

At Giggleswick Scar, North Yorkshire, the Craven Fault system
has left Great Scar Limestone high above the Bowland Shales of
the plain below.

Volcanic Scenery

Fortunately, there are no active volcanoes
in the British Isles and it is several million
years since any vulcanism occurred. However, the Tertiary volcanic period, and
previous periods of similar activity, have
left us a rich legacy of volcanic study areas.
Tertiary vulcanism was probably prompted
as a consequence of the Alpine period of
earth movements which displaced and
cracked some of the older rocks of north-
west Britain, thereby producing fissures
through which magma could reach the
surface. The main areas affected were the
Hebrides and northern Ireland and, as we
have seen, much of the areas around and
between. Dykes thought to be related to
the centres of activity, on Arran, Mull,
Ardnamurchan and Skye, are intruded into
strata as far south as northern England.

The coastlines of Morvern, Mull and
many of the Hebridean islands reveal sections of the many recurring flows of lava.
Natural cathedrals, caves and spires have
been sculpted from the hexagonal columns
of basalt which, in some districts, are
sufficiently soft and crumbly to break down
into a fertile soil; particularly where the
lava contains vesicles of calcium carbonate
and related zeolites. Fertile soils and the
mildness of climate induced by the Gulf

Stream have produced a verdant landscape which is typified by the emerald green lands of Co. Antrim, near the Giant's Causeway.

In Western Mull the columnar lava is inter-bedded with the casts of fossil trees which were encapsulated in the engulfing molten lava at Rudha nah-Uamha. Otherwise there is a sparsity of such finds.

Many of the lava flows have built up into a layered pseudosedimentary structure. Harder, more acid lavas give some of the valleys of Mull a stepped almost Yoredalian aspect. The cliffs of the island of Canna have been eroded by the sea to reveal sections of basalt lavas inter-bedded with doleritic sills, tuffs and agglomerates from vental explosions; all resting on a base of pre-Cambrian conglomerates. Perhaps nowhere else in the British Tertiary volcanic districts is the stepped nature of successive lava flows better exposed than on the slopes of Compass Hill on Canna, so-called because its high iron content makes navigation around Canna somewhat hazardous.

Travelling back in time to the Carboniferous period, there was an earlier period of volcanic activity instigated by the eruption of central vent volcanoes. These were responsible for lava flows and ash sheets of a more restricted nature. Several of these Carboniferous volcanoes became choked with lava which solidified into plugs of basalt. Edinburgh Castle stands on Castle Rock, a plug of basalt lava which was isolated from the neighbouring landscapes by the eroding action of ice sheets on the surrounding, softer sedimentary rocks of what is now the Waverley (Gorge) railway station and gardens. Nearby, Arthur's Seat is a conglomeration of several volcanic necks. The east coast gannet sanctuary of Bass Rock provides yet another example of the durability and hardness of a basalt, volcanic neck that has defied the onslaught

Edinburgh Castle stands on a volcanic crag.

of the North Sea for millions of years.

South of the Scottish border, the Whin Sill, a quartz dolerite sill of Carboniferous age, underlies much of northern England, stretching from Cumbria to Northumberland and, as evidence from drilling has proved, continuing far beneath the North Sea. Its northern outcrop provided a firm craggy base for Hadrian's Wall to the east of the Tyne Gap at Cawfield Crags and Whinshield Crags. At its western-most outcrop, near Cross Fell, sombre cliffs of the sill stand out in contrast to the white Carboniferous limestones above and beneath it. In various places the Whin Sill has been quarried for its wear-resistant roadstones. The sill's hardness has been a contributory factor in the development of waterfalls at Cauldron Snout and at High Force, where the hard sill has protected the softer sedimentaries below from erosion and cutback by the River Tees. The thickness of the Whin Sill varies from 73m at Burtree Pasture lead mine in Weardale to as little as 2m in places where it has divided into several leaves. At one time the Whin Sill was thought to be a lava flow interbedded with, rather than injected into, the surrounding strata.

In Wales the problems are more complex as many sills and lava flows are found within strata of Upper Cambrian to Lower Silurian times. The Cader Idris scarp face is formed predominantly of granophyre, which was probably intruded at the same time as some of the Ordovician lava sheets were extruded.

Granites

Granites play an important role in the structure and mineralisation pattern of Cornwall, as they also seem to do in other parts of the country. Cornwall is underlain by a massive granite batholith stretching along the peninsula from Dartmoor towards Scilly. Four large protrusions (bosses) of this granite, which was intruded about 275 million years ago, have been exposed by erosion. These are found at Land's End, St Austell, Carnmenellis and Bodmin Moor. Smaller exposures of granite outcrop in line with, and between, the larger bosses. St Michael's Mount in Mount's Bay is a popular example. Granites lie beneath the Lake District dome and their hard yet fracturable structure is responsible for the stark crags and screes between Wastwater and Ennerdale.

Granites are most abundant north of the border and north of the Highland Boundary Fault, along the Grampian mountains and the Spey valley, where the river has exposed the intermingled schists and granites which lie at the heart of the rugged, windswept Cairngorms. The pink Cairngorm granite has been freshly exposed by quarrying at the foot of the road to the Cairngorm ski lift. In its pristine freshness it is one of our most colourful granites, rivalling Shap granite in aesthetic qualities. This pink and white, coarsely crystalline granite adds a touch of colour to the blue-black drabness of the high Cairngorms. (Translated from the Gaelic the name means blue mountains.) The granite is very coarse because it was intruded at great depth into the surrounding Moine schists and, therefore, took a long time to cool. As cooling occurred there was, as a result, some shrinkage which made cavities into which yellow and smoky quartz were intruded. Tourists buy these trophies as 'Cairngorm' gemstones. Further east the grey granites of Aberdeen outcrop, lending an air of drabness yet durable dependability to the area.

Although granites are important as deep-seated intrusives which support the rocks above them, their outcrops at the surface – the tors of Dartmoor, the Cairngorms and elsewhere – are only a small representation of the true scale and amount of granite in the crust. The fact that granite is a deep-seated intrusive means that it can be exposed only after many years of uplift and erosion. Thus, the older the land, the greater the amount of granite exposure.

When granites and other igneous rocks are intruded into sedimentary strata some alteration of the surrounding rocks occurs. Metamorphism is also produced by folding and compression. The distinctive scenery of Wales, part of the Lake District and much of the Scottish Highlands is a result of metamorphic processes.

Slates

All three areas mentioned above have been extensively quarried for slate, and the hillsides of Llanberis in Wales are pockmarked and littered with the remains of past industry. Slates are hard and impermeable, so the slate hills and mountains of Caernarfonshire are badly drained and swampy in hollows. The slate deposits reach a maximum thickness of 900 metres near Bethesda and about 700 metres at Llanberis. Colours vary from green to purple, but it is the purple kind which predominates.

Cumbria is more noted for green slate, although blue and grey varieties are also

Unconformity, Thornton Force, Ingleton, North Yorkshire. The figures are at the junction of Lower Palaeozoic Ingletonian strata and horizontal Carboniferous limestones above.

found. Skiddaw and nearby mountains are formed of a hard grey Skiddaw slate which defies erosion and remains as steep ridge and corrie country beloved of mountaineers. Skiddaw itself is a dull climb, rather like the ascent of a coal-mine spoil heap, and the windswept, infertile, slatey summit provides little in the way of greenery for the eye that has been jaded by the monochrome grey of the slate all the way up. Further south, the blue-grey slates of Coniston and the green slates of Tilberthwaite Gill lend a much more colourful background to the woodland and mountain pastures that struggle for a hold on the thin, acidic soils of the area. These districts are noted for single span slate bridges, called clapper bridges by local people.

Slates and greywackes of a similar early Palaeozoic age are found a few miles to the south-east of the Lake District, in the Ingleton and Clapham areas, where several inliers of older rock have been exposed by uplift and recent erosion by rivers. Ingletonian strata of possible pre-Cambrian age provide some of the best waterfall scenery to be found in Britain. A walk up the valley of the Twiss through glens of broad-leaved woodland, past cascades formed where hard outcrops of slate have resisted erosion, is an instructive geological excursion. At the top of the walk, Thornton Force plunges 13 metres from the rim of a natural amphitheatre into a circular pool below. Under the fall the old rocks meet the Carboniferous limestones in an unconformity that represents a time gap of at least 300 million years.

The slate districts of Scotland are situated on the west coast, at the bottom of Glencoe at Ballacuilish and on the islands of Seil and Luing. A small island off the coastline at this point, Belnahua (Island of Slate) was quarried so extensively that the sea flooded in to the centre of the island. Now the hollowed and worked out central hill of Belnahua stands exposed, like the rib-cage of a stranded whale. This Scottish slate is dark and dull like the scenery around it, but the picturesque whitewashed cottages (once the homes of quarrymen) at Easedale add life to the scene. Much of the Scottish slate contains cubes of pyrite which weather to rust, leaving holes in the roof. Not surprisingly, the industry is long abandoned although it once had the distinction of providing slates for the roof of Iona Abbey.

Slates and related metamorphic rocks, schists and gneisses provide much of the harsh but splendid mountain scenery of the Highlands. These areas have been so gouged by glaciers, flooded by rivers and the sea that transport is difficult and man has had little chance to utilise the land as in the more southerly regions of Britain.

6 Man and Geology

Beginnings

Man's progress from prehistoric times to the present has been inextricably linked with his understanding of the natural world and the discovery of its resources. In some cases by accident, in other instances because an individual had the superior intellect to adapt a raw material into a useful product, we have learned how to utilise the minerals and rocks with which we are abundantly provided. It could be argued that the knowledge of geology has been the major factor in our development. Certainly, we have made use of a rudimentary knowledge of geology at least from the time our primitive ancestors began to use tools and live in caves.

Caves, for instance, occur most abundantly in limestone areas, which also tend to be more fertile and, therefore, able to support a greater variety of food plants and animals. It was no accident that greater concentrations of early hominids tended to accumulate in limestone districts where natural shelter could be found. Chalk districts, too, were favoured. Chalk could be excavated into caves and, more importantly, provided the raw material, flint, which became perhaps the first commercially exploited mineral resource at locations like Grimes Graves, a Neolithic flint mine at Weeting, Norfolk.

Flint is a variety of silica. Found as

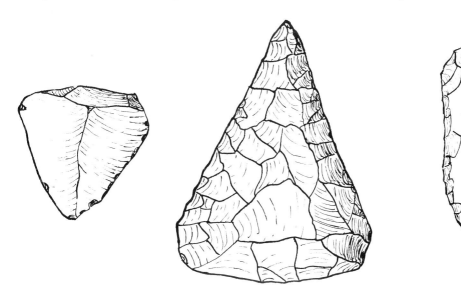

Flint, a silicate, fractures conchoidally to a sharp edge. It was one of the first natural resources to be excavated.

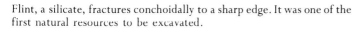

105

nodules, in a variety of contorted shapes, flint is hard and breaks with a glass-like, conchoidal fracture into sharp points and edges. Long before man began to work flint, pebbles of various shapes resembling tools and axes may have been used as found. Some are so crudely worked that it is only their proximity to known dwelling sites that proves their utilitarian purpose. Others known as eoliths, were avidly collected by Victorian archaeologists. Eoliths are now thought to be naturally shaped flints, although it is very difficult to decide whether some flints are naturally formed or were shaped by an inexpert, early flint knapper. Similar siliceous materials, chert and obsidian (a glassy volcanic lava), have also been used as tools.

In order to understand the difficulty of working such an intractable material as flint, it is a useful exercise to attempt to break a piece into a dagger or arrowhead – without cheating by using modern tools; but using only other flints or stones. Flint seems to break anywhere but where it is intended to break. After much experiment it becomes possible to discern probable lines of fracture, but it takes years for any modern man to achieve the precision of some of the later flints which were still in production at the beginning of the Bronze Age. (Note If you want to try this experiment you should wear a pair of safety goggles.)

Cave Paintings

At the same time as man was acquiring manual dexterity by working flint, his hand and eye co-ordination had reached such a stage of advancement that he was able to depict scenes of hunting, everyday life and religious symbols in cave paintings. Pigments were mixed with fat, urine or blood to give the required consistency. Iron oxides provided red (hematite) and yellow (ochre) paints. Mineral samples were ground in hollow stones (and here we find another use of stone) to a fine powder, before being mixed with the appropriate base. Later developments were the use of green and blue copper carbonates and white zinc carbonates as pigments.

In order to grind paints and also grain, querns were fashioned from suitable stones. Originally, any hollow stone could have been used, but later the stones were quarried out of coarse sandstone or gritstone found at localities like the appropriately named Quernmore in Bowland, Lancashire. Man was slowly learning how to quarry and work stone. Much later, the same types of gritstones (millstone grits) would be used to make millstones.

Cave artists did not, of course, work in the dark. Suitable stones were carved into elongated saucer-shaped lamps. Many small utensils were made of stone. These were so durable that they lasted for generations, but they were crude and clumsy and by Neolithic times were beginning to be superseded by pottery items.

It is thought that women were the first potters. Possibly, they noticed that a piece of soft clay had become hardened in a cooking fire; or perhaps clay was used in the way that gypsies cooked hedgehogs, as a disposable protective crust. In whichever way the idea was conceived, pottery making began with primitive thumb pots and soon advanced to coiled pots, then – using some form of primitive turntable – to turned pots. Suitable clays were found almost everywhere, although skilled potters tended to live near to the finest sources, and trade routes built up along the trackways leading to the sites of these resources.

The Bronze Age

Flint provides a sharp but fragile cutting edge on tools and weapons. At the end of Neolithic times, 6,000 years ago, bronze was the new wonder material because it was more easily worked and not as likely to fracture as flint.

The Bronze Age began at different times in various parts of the world. The oldest known bronze objects were found in the Shanidar area of Iraq and date back over 10,000 years to the time when agricultural practices were first developed. In Western Europe we regard the bronze working process as having been developed about 6,000 years ago.

Bronze does not occur naturally and it is likely that the first metal workers used native copper (copper found in its metallic state, not as an ore). No doubt this malleable metal was worked cold until someone found it to be more easily shaped when heated. The next step was to try heating the more brittle copper ores that were found with native copper to see if they too could be made more workable. Thus the smelting process was invented. Chalcopyrite and other copper ores are rarely found in isolation from other minerals and in some areas there are transitional stages of mineralisation where copper and tin are found together. By smelting these 'impure' copper ores a golden yellow metal (bronze) was probably produced accidentally. A fortuitous accident, for this new metal alloy was harder and gave a more durable cutting edge than copper. Analysis of early bronze shows that its tin content ranged between five to fifteen per cent. Later, it settled at about ten per cent, as bronze workers began to smelt to a formula.

Although copper was commonly found over much of Europe and in the Middle East, tin was less abundant. Large quantities were found only in Egypt and, to a greater extent, in Cornwall. Smaller amounts were discovered in Europe, but the Malaysian deposits were not utilised until much later by nineteenth-century European traders. The discovery of bronze instigated the large-scale expansion of the mining industry, and Britain became an important source of tin.

Cassiterite or 'tinstone' was exported from Cornwall, by sea, to northern France. Thence it was carried to Marseilles through the Seine and Rhône valleys. From Marseille, tin could be transported by sea to countries bordering the Mediterranean.

At first, much of the tinstone was dug from alluvial deposits in the beds of streams where it had lodged because of its high specific gravity. As stream tin sources began to dwindle, the veins from which it had been eroded were located and proper mining began. The tin ores were found within and close to the intrusions of Cornish granite, so the construction of tunnels though the hard rock was slow and arduous work. It is difficult to say just how long the Cornish tin trade lasted, but certainly the discovery of precious metals and tin at Tarshish (Cadiz), recorded by Ezekiel, hastened the decline of the Cornish industry which remained dormant for many years.

The Iron Age

By about 500BC, mining and smelting techniques had progressed to such an extent that men were able to produce the high temperatures that were necessary to make iron from hematite ore. The product was a grey, spongy slag which soon scaled and rusted, but smiths learned that by heating and hammering the raw metal it could be formed into crude shapes. At first, iron was used only for small ornaments, but later it was to usurp bronze as the utilitarian metal.

Hematite was extracted from the Furness district of Cumbria and to a lesser extent from other areas of Britain. Iron meteorite pieces, which were occasionally found, were highly prized. These were smelted and made into hard, razor-sharp weapons which, with the benefit of modern analytical techniques, have been proved to contain nickel and cobalt and are, therefore, a type of raw steel.

Iron is such a cheap and useful metal that it has been called the democratic metal – a metal affordable by all. The Iron Age has continued up to the present day and the metal still forms the basis of many mass-produced products.

The Romans in Britain

By the time the Romans came to Britain, the tin trade had declined to such an extent that a report by Julius Caesar makes only slight mention of the metal and wrongly attributes its source to the interior of the island. Much more important to the Romans were the rich deposits of lead which they used in drinking vessels and plumbing systems. Pigs of lead bearing Roman inscriptions have been found at various locations in the Pennines, and Roman mining implements and artefacts have been discovered in deep levels of modern mines. Furthermore, the Romans were aware that lead ore often contains small amounts of silver. They were such skilled smelters that they were able to extract silver from lead ore which contained only a few ounces of silver per ton. Their road systems often veer towards known lead mines and it is thought that some of their camps, like that at Bainbridge in Wensleydale, were centres for lead trading.

In the construction of roads and buildings they used local stone which was quarried nearby. Roman buildings were held together by a cement which, it has often been said, has remained without equal to the present day. This was true until Victorian times when the secret of cement making was rediscovered and the construction of large concrete buildings began. Roman cement was made from clay and limestone or from volcanic ash and slaked lime. It has withstood the test of time and many Roman constructions are still standing.

After the end of the Roman occupation of Britain there was a period of retrogression or stagnation when it is unlikely that the extraction of minerals and stone supplied anything other than basic necessities. However, the skills endured and from the Middle Ages onwards our extractive industries prospered. Tin, iron and lead mining were revitalised and continued to thrive until cheap foreign imports brought about their demise. The invention of steam power brought new industrial methods and more uses for an iron trade which really took off and spread from Coalbrookdale, Shropshire. At the same time less wood and more bricks and stone were used by the construction trade.

Britain was richly provided with all the coal necessary to fire the processes of the Industrial Revolution. As better steam engines were developed, the mines could be pumped dry more efficiently and even more supplies of metallic ores and coal were then available to satisfy the insatiable appetite of industry.

The Extractive Industries

Limestone

Limestone is a durable building stone, but its most important use has been as a land improver. In districts where limestone strata occur, or where limestone boulders could be found in glacial drift deposits, the stone was quarried or collected to be made into 'quicklime'. Farmers burned the limestone with coal, peat or wood in a specially constructed kiln. After firing, the quicklime (calcium oxide) was raked out and slaked with water to produce slaked lime (calcium hydroxide) which, when spread over the land, acted as a fertiliser. Liming

Before the advent of large scale quarrying, limestone was burned in small coal-fired kilns to produce lime for agricultural purposes.

improves land by counteracting a natural build-up of acidity from humus deposits. Thus, more minerals are available to assist healthy plant growth.

In 1824 Joseph Aspdin invented a calcined powder of limestone and clay which he called Portland cement because, when mixed with water and allowed to harden, it resembled stone from the famous Portland quarries in Dorset. The process has been improved so that now the mixture is heated just to the point of fusion then ground down to a powder.

Clay materials can be obtained by grinding down shale, slate, furnace slag or sand. Nowadays, about three per cent gypsum ($CaSO_4$) is added to the limestone and argillacious material. Modern variations on cement include alumina cement, which is a mixture of roughly equal parts limestone and bauxite (aluminium ore), fired to the molten state, cast into pigs and ground to a powder. Various aggregates such as sand, gravel, granite chippings, and so on, are added to make concrete.

A further use for limestone is in glass making. The main constituents of glass – limestone soda and sand – are heated to a molten state. The hot fluid then passes to a 'float bath' where it is supported on a bed of molten tin whilst heat is applied from above to keep it flowing evenly. As the glass passes over the float tank the heat is progressively reduced until the glass is sufficiently hard to be fed into the rollers of an annealing oven without marking. After annealing the glass is cut into lengths.

The demands on limestone are considerable and there are quarries in every district where good supplies are found. Unfortunately, many quarries are situated in areas of natural beauty and are often thought to be an eyesore. To the geologist, a limestone quarry is always an interesting

Quarrying in Carboniferous limestone.

feature, an economic necessity and also a good hunting ground for fossils. Without a limestone quarrying industry there would be no cement to hold our houses together, no concrete for civil engineering projects, no lime for the farmers, no hardcore for roads and railways, no glasses to drink from or windows to light our homes, no spectacle lenses, telescopes or cameras, and no steel (limestone is used in blast furnaces). In fact, it is doubtful if we would have much of a civilisation at all without limestone.

Brickmaking

Brickmaking, a traditional craft dating back thousands of years, also relies on a good supply of raw materials. An ordinary brick is made of any suitable clay material fired to

a sufficiently high temperature to fuse the individual particles together. A suitable material may be found as shale, surface clay (boulder clay) or fire clay from under a coal seam. Whatever the source, a brick clay must be plastic enough to mould when ground down and mixed with water, yet possess sufficient tensile strength to retain its shape after moulding. The steps in brickmaking are: mining or quarrying raw material; preparing and shaping; drying; firing; cooling.

The chemical composition of a brick clay determines its final colour. A basic clay consists of silica and alumina with small amounts of metallic oxides and other minerals. Calcareous clays contain some lime, and fire to a yellow-coloured brick – the favourite of the old Midland Railway for the building and decoration of stations and houses. Iron oxide impurities produce buff, red or pink bricks. The best bricks for refractory linings are made from mineral-deficient fire clays.

Plaster

Another building material for which it would be difficult to find a substitute if supplies become depleted is plaster, made from gypsum, an evaporite deposit. (The Greeks and Romans used gypsum in many of their structures.) In Britain the main source of gypsum is the Permian evaporite deposits in the Eden valley south of Carlisle.

To make a basic plaster of Paris, ground gypsum is heated to a temperature of about 330° F. This drives off about 75 per cent of the water which is usually combined with the natural mineral ($CaSO_4(2H_2O)$). Anyone who has suffered a broken limb will realise the uses to which plaster of Paris can be put. It is also used with a variety of additives to make wall plaster, plaster boards and general purpose fillers.

Quarrying

Before the age of quick transport, builders were usually limited to using local materials. Every district had a brickworks or quarry and it seems, from an aesthetic consideration, that local materials blend best into the landscape. Although some larger, more important structures were built of imported materials, it was not until the spread of railways that builders were able to construct with unsuitable materials that, even when mellowed, looked out of place.

Welsh slate roofs replaced original thatch or local slates and tiles in most parts of Britain. The beginning of the railway era brought about a revolution in the churchyards of Britain during Victorian times. Now green slate headstones stand next to marble angels, sandstone crosses flank granite obelisks in a multi-coloured hotchpotch of monumental masonry. Architects were unable to resist their new opportunities to decorate public buildings with alien stones. Aberdeen and Shap granite facings were applied to the facades of banks. Town halls were given colonnades of marble columns. Dark Scandinavian laurvikite was used wherever a dignified but decorative appearance was required.

It was during Victorian times that quarrying became an international business. Sometimes the products of quarrying were available so cheaply that they were used as ballast by ships which had imported cargoes to Britain. In this way sandstones from north-west Britain were taken to the southern states of the United States by ships which had brought cotton to Liverpool. Buildings made of British sandstone still adorn the streets of the cotton towns of the deep South.

The construction of railways made heavy demands on the quarries. Ballast was needed to provide a solid foundation for rails and sleepers. At first, unsuitable or broken stone could be used as a by-product of quarrying for building stone. Later, some quarries were adapted to the exclusive production of broken or crushed stone.

Town centres were paved with solid blocks of sandstone or granite sets and country roads were surfaced with crushed stone. Limestone was a favourite material for this purpose. Eventually, the process of tarring finely graded stones and compacting them into a hard surface made roads more suitable for motor vehicles. Modern quarries use all the aids of computer technology to ensure uniform hardness, durability and

Geology in the town. The Halifax Building Society is faced with
sandstone and pink granite. The Yorkshire Bank is faced with limestone
and grey granite.

Computerised road-stone quarry, Helwith Bridge, North Yorkshire.

size of the chippings produced for road building. Greywacke and granite chippings are much favoured for this purpose because of the anti-skid properties of the former and the hardness of the latter.

Metals and Mining

As all forms of transport became faster and more efficient, ever-increasing quantities of metals began to be used in the construction of ships, cars, lorries, bridges and aircraft. Advances in food technology brought about the introduction of the first convenience foods when tin-plated steel cans were used to preserve food. Tin mining had continued in Cornwall from prehistoric times up to the nineteenth century. The increased use of tin brought about a renaissance of the mining industry in Cornwall as miners delved ever deeper for the rare cassiterite. British lead mining experienced a similar surge of production.

As industries demanded ever-increasing supplies of metals, prices soared. Miners sought minerals abroad and such vast deposits were found that prices fell, the British mines became uneconomic and by the beginning of the twentieth century most of the mines were closed. The hematite trade in Cumbria continued until new smelting techniques were invented that were capable of using inferior, foreign iron ores.

In modern Britain there are few metallic ore mines, although potash, barite and fluorspar continue to be mined profitably. However, there are signs of a reversal in the trend. International mining companies like Rio Tinto Zinc (zinc accounts for only a small percentage of their products) attemp-

Old trucks once used to carry lead ore from the mines at Leadhills.

ted to reopen the old Cornish tin mines. Prospecting for rarer metals is taking place in Scotland. Gold has been found in Northern Ireland, and Europe's largest lead zinc mine was successfully prospected and constructed at Navan in Co. Meath, Eire. The likelihood of a decline in the availability of minerals has spurred on this renewed activity, and it is quite possible that previously unexploited mineral deposits may be obtained at a much greater depth than erstwhile mining technology was able to reach.

Deep mining techniques were developed in South Africa in order to follow the dipping gold reefs of the southern Transvaal and Orange Free State. As shafts were driven down towards the furnace of the earth's core, rock temperatures of 45°C were found at depths of 1,700 metres in the Free State and at 2,200 metres in the Transvaal. In the future it is envisaged that mines may penetrate to depths of five kilometres where temperatures around 80°C may be expected. How can man, with a skin temperature of 35°C, survive at these mephitic temperatures? The answer lies partly in man's adaptability to working in high temperatures after a period of acclimatisation, but more particularly in the installation of air refrigeration equipment and cold water sprays directed at exposed rock faces. No British mines have ever reached such depths, but they may do so in the near future. Although subterranean heat problems are not yet encountered in the British Isles, the average geothermal gradient is about 25°C/km. However, in districts of above average geothermal gradient, temperatures as high as 70°C may be anticipated at depths of two kilometres.

Temperature control is not a high priority problem in the shallow mines of the British Isles. One of the major concerns in a region of high population is what to do with the spoil or gangue material (everything that remains after the metal ore has been extracted from the material brought out of the mine). Tara Mine at Navan provides an excellent example of good design and planning which enables even a large mine to be sited very near to a town without incurring environmental problems. To anyone except devotees of mining, geology or the flora of spoil, the waste heaps of many an old mine appear to be an unnecessary blot on the landscape. At Tara there is little to indicate that a working mine exists. To the casual observer the workings would appear to be a tastefully

Exploration work to re-open an abandoned mine adit.

designed, well-landscaped factory. Environmental controls cost the Tara mines company $10 million to incorporate into their workings.

The discovery of the major lead zinc ore body at Navan illustrates how geologists use the most up-to-date technology to find ore which may be hidden beneath drift deposits and barren strata. Before the first borehole at Navan was started in 1970 the company had been prospecting in Ireland for eight years at a cost of $1.5 million. During this time 148 diamond drillholes had been sunk over 33 of the company's 165 licensed prospecting areas. From 1968 to 1969 licences were obtained near Navan in an area underlain by lower Carboniferous limestone. Although there were no surface indications of ore, analysis of soil samples showed a high regional zinc anomaly. (This type of prospecting is known as geochemical prospecting.) Induced polarisation techniques were used to define suitable sites for drilling. Electrodes were inserted into the ground, a unidirectional current was applied for a short time then switched off. A fraction of a second after switching off the source of power, the potential difference betwen the electrodes was measured. Induced polarisation methods rely on the fact that large bodies of metallic ores retain some measurable electrical charge for a short time.

Once a rough idea of the location of the ore had been obtained, boreholes were drilled and 38mm-diameter core samples were brought to the surface for assaying. From analysis of these core samples it was determined that high grade sphalerite and galena deposits occupied fractures near the base of the Carboniferous. Mineralisation had fanned out from the fissures and replaced much of the host rock. The orebody was found to contain an estimated 61 million tons of ore grading 11 per cent zinc and 2.4 per cent lead.

Not all these reserves became available to the Tara Exploration Company, as a rival company had bought leases of 15 per cent of the ore north of the Blackwater river. However, it was worth continuing and development shafts were sunk. Then a production shaft and declines at a gradient of 20 per cent were constructed. This allowed access for the 26-ton haulage trucks and other mine vehicles. Transport of personnel was also via the incline.

To avoid the massive spoil heaps that are often associated with mines, the method of working Tara is by the cut and fill method

Equipment used for shallow, exploratory drilling.

Loading ore at Tara Mine, Europe's largest lead-zinc mine.

whereby most worked-out areas are refilled with a mixture of crushed spoil and cement.

Other environmental considerations included pre-development baseline studies of the chemical qualities of the Boyne and Blackwater rivers, their biology and hydrology; soil and vegetation surveys; veterinary survey; drinking water analysis; atmospheric survey; sound and vibration surveys.

To maintain the standards found by the surveys, landscaping, tree planting (120,000 trees and shrubs of 70 different species), and process water treatment plants were installed. On top of all this the management maintain an environmental monitoring programme in order to avoid the possibility of any form of pollution. The entire enterprise shows that resource industries can supply much-needed raw materials without damaging the environment. The non-miners in the nearby town of Navan hardly notice the existence of Tara Mine except for the fact that many jobs and the prospect of future prosperity have been made possible by the skill of the geologist.

That, indeed, is a primary purpose of geology – to provide the raw materials without which civilisation would regress. Gold for jewellery and, perhaps more importantly, for electrical connections in silicon chips. Platinum, the multi-purpose catalyst that remains unchanged by chemical reactions. Tungsten for lamps, aluminium for aircraft, iron for ships; the list is endless and inexhaustible, but we must add uranium – the energy resource of the future – and the fossil fuels – coal, oil and gas. By 'fossil' fuels we mean energy supplies that have been locked in the earth as a result of

116

Multiple shot hole drilling at Tara Mine.

Process water ponds, Strontian barite mine, Argyll.

sedimentation over decayed remains of living things. The gas fields may be the products of putrefaction of billions of strata-locked creatures. Oil is derived from the remains of their fatty parts.

Oil

Crude oil, the black, treacly substance that comes out of oil wells, is a mixture of various hydrocarbon molecules from dense tars to ethane and methane. The whole mixture is separated into various petroleum products – petrol, tar, paraffin, diesel, and so on – by cracking (or distillation) at the refinery. As the use of petroleum products grew with the development of the internal combustion engine and the needs of industry, there have been several periods of scarcity which were, if we listen to the pessimists, indications that the world oil reserves are dwindling. So far the oil exploration geologists have always managed to find new oil fields at greater depths or in previously unexplored areas. The North Sea was a typical case, and it is a distinct possibility that further finds will continue to be made in geologically suitable areas around Britain's coastline and on land as well. From scarcity, the pendulum usually seems to swing towards glut, due to rising prices and added financial incentives to discover new fields or bring into production previously uneconomic fields.

If the majority of geologists are correct, there must be an end to the oil reserves even though it may take hundreds of years to exhaust the supply at projected consumption levels. On the other hand, there is at least one scientist who is opposed to the accepted theories of the biological origin of oil. Professor Thomas Gold of Cornell University, New York, believes that hydrocarbons, gases and oils are constantly being produced in the deep interior of our planet.

As an astrophysicist he points out that spectrographic analysis of the outer planets and comets shows that methane and other hydrocarbons are present in vast quantities. Yet it is highly unlikely that any form of biological action occurred before the surfaces of these planets cooled and froze. Even the presence of biological molecules in petroleum cannot be regarded as positive proof of its biological origin. The molecules may have been absorbed as the petroleum seeped through sedimentary rocks or biological alteration could have happened after formation. We know that carbonaceous material exists at depth because diamonds (a very pure form of carbon) are formed at great depths. Carbon dioxide is produced by volcanic eruptions, so this may be an indication of a deep-seated origin.

The only way positively to prove these theories is by deep drilling. Such a project is under way in Sweden at a point where a gigantic meteorite impact has shattered local granites and methane seeps to the surface at a number of sites. Here there are no sedimentary rocks, so any hydrocarbons found cannot be a product of sedimentation. If a hydrocarbon reservoir is found our understanding of hydrocarbon formation and our estimation of world energy reserves must change.

Hydrocarbons are found by well tried and tested geological practices which rely on the fact that oil floats on water and migrates upwards through porous strata until it can rise no further and is trapped beneath a cap rock. In some areas of the world seepages indicate that oil has passed to the surface through faults or fissures and that larger supplies may be obtainable beneath. However, in most districts, there

(a) Anticlinal trap

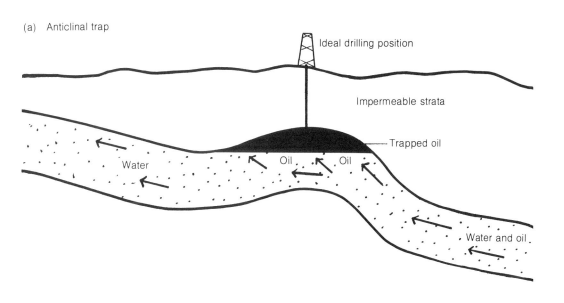

(b) Fault trap

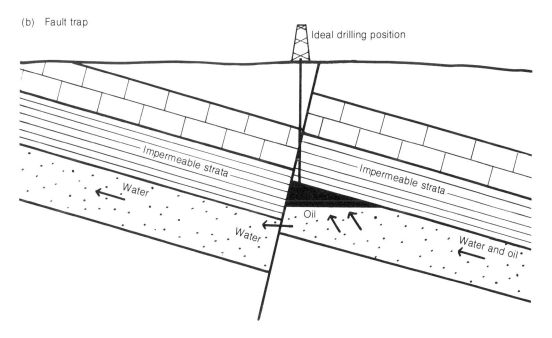

Entrapment of oil by migration of water and oil through permeable
strata capped by impermeable strata.

are no traces of oil at the surface and recourse must be made to geophysical prospecting techniques, guided by a knowledge of structural conditions.

One of the most important structures to petroleum geologists is the anticline, which may form an ideal structural trap along its crest, particularly if the crest is surmounted by impermeable strata. Other suitable traps may be located where faulting has juxtaposed dipping permeable and impermeable strata, with the impermeable rock sealing the faulted surface of the permeable host rock which is also overlain by the impermeable rock. Interpretation of this type of feature needs not only a knowledge of faulting but also a mind able to cope with the three-dimensional prediction of what lies below the surface. Anticlines may exist at depth even though their presence is barely apparent at the surface. Such anticlines may unconformably overlie an eroded basement of hills and valleys. Sediments accumulate more thickly in valleys and a previously hilly basement area may build up to a level plane, underlain by layers of sediments which formed a level surface before compaction and only a slightly folded surface topography after compaction. This type of structure also forms a suitable anticlinal trap for oil. Sand bars and reefs also impede the migration of oil, as do unconformities with impermeable strata situated above the plane of unconformity.

The methods used to detect suitable oil bearing structures include gravity surveying and airborne magnetometer surveying. The former is used to detect granite plutons which would cause domal formations where oil could accumulate. The latter method may detect occurrences of high magnetic basement rocks which would also cause doming.

Perhaps the most important geophysical technique for detecting oil is seismic surveying which employs methods similar to those used in the study of earthquakes, except that the explosive pulses are manmade. Basically, shots are fired in shallow boreholes, or, alternatively, mechanical vibrations are emitted from specially adapted vehicles, and the shock waves produced are detected by a complex arrangement of geophones. The success of seismic surveying relies on the fact that different types of rock have different sound absorption and refraction patterns and a picture may be built up of the internal structure in a way that is similar to the detection of underwater objects by sonar.

All oil detection processes may give misleading evidence and the only way to prove or disprove the presence of oil is to drill. The South China Seas contain many suitable structures but drilling in several promising areas proved that most were dry or contained oil in uneconomic quantities. The British Isles have been more fortunate and there are possibilities of new land-based oilfields but, until the reserves are proved, the old adage, 'where you find it, there it is', is the nearest we can get to accurate prediction.

The science of rotary drilling began in 1818 when the corporation of Paris sank a borehole for water. Development of the equipment and methods progressed throughout the century, but it was not until the 1930s in Texas and Oklahoma that rotary drilling was fully exploited. The basic method is simple enough; a diamond impregnated drilling bit supported at the end of hollow iron rods is rotated in the borehole. A drilling mud which acts as lubricant, coolant, scourer and cleaner is pumped down to the bits through the hollow drill rods. Waste materials pass to the surface between the walls of the

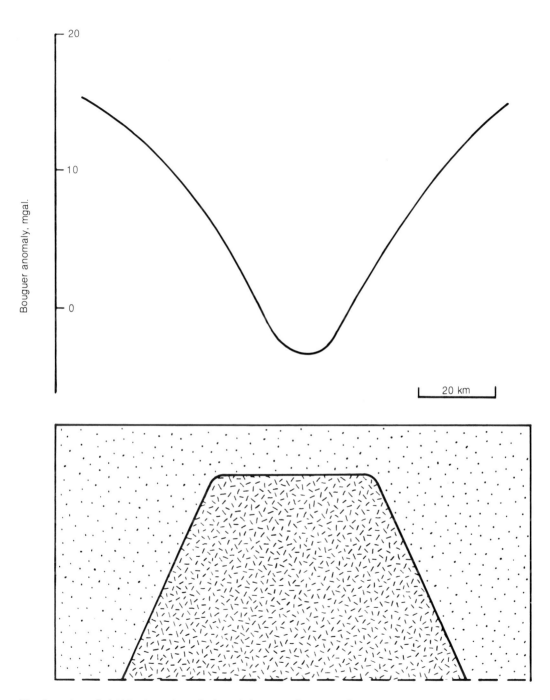

The detection of a hidden intrusive, of relatively low specific gravity, by gravity surveying. Note the negative anomaly curve on the graph.

A seismic survey in the Craven Basin. A plate beneath each vehicle contacts the road and emits vibrations which are reflected by deep structures and picked up by sensitive geophones.

borehole and the drill rods. When the drill bit has penetrated about ten metres, another drill rod (of similar length) is screwed on to the column of rods (constructed like a chimney sweep's extendable brush) and work begins again. As the borehole gets deeper the column of rods grows longer until the backlash effect of the extended column is phenomenal. Despite such difficulties, Russian drilling teams have penetrated to depths of twelve kilometres.

Accurate drilling techniques are not only important in the detection of oil and minerals, but also help our understanding of the earth's structure, and are a valuable aid in discovering what goes on beneath the surface. As we exhaust our finite energy resources other ways of obtaining energy will have to be exploited. In areas of the world where volcanic activity takes place, high temperatures are often encountered very near to the surface. Geysers and hot

A drilling rig on Furzey Island, Dorset.

springs may be harnessed to provide geothermal energy resources for home heating and to provide a free source of heat for growing vegetables. Iceland is the country best known for these practices, but Britain may soon join the list of geothermal energy users. Commercial enterprises have been formed to drill and test the possibilities of pumping water down to hot granites and back to the surface via a nearby return borehole. The deep granite, warmed by the decay of radioactive minerals, could provide enough heat to warm thousands of homes. The only cost entailed would be for drilling, pumping and servicing of equipment.

Further research has been carried out by the Geological Survey who have drilled boreholes along a belt of country stretching from mid-southern Scotland across the Pennines to the Wash. Here, geothermal heat flow is slightly above average. There are many possibilities of tapping the earth's geothermal energy, not least where faults act as channels for deep-seated thermal springs to reach the surface.

The Future

Much remains to be discovered in the science of geology. Although advances in our understanding of the earth are constantly being made, we are always working in the dark because everything is blanketed by a cover of rock. What is so fascinating about geology is that all our efforts at drilling, investigating and sampling are man's last venture into the unknown in a world that has otherwise been explored and mapped. Slowly, however, the jigsaw of geological knowledge is being assembled and we are beginning to understand some of the complexities of the earth's structure and its past development. Even so, there are still some areas in Britain which have not been surveyed in detail and there are many areas of the world yet to be thoroughly geologically explored and mapped.

Further Reading

Bateman, A.M., *Economic Mineral Deposits* (Chapman & Hall, 1952)

Barton, R.M., *Geology of Cornwall* (D. Bradford Barton, 1969)

British Geological Survey, Memoirs to the 1 inch and 1:50,000 sheets (B.G.S.)

Deer, W.A., Howie, R.A. & Zussman, J., *An Introduction to the Rock-Forming Minerals* (Longman, 1969)

Harris, A., *Cumberland Iron* (D. Bradford Barton, 1970)

Hedges, E.S., *Tin in Social and Economic History* (Arnold, 1964)

Hickling, G. (Ed.) *Geology of Durham County* (Natural History Society of Northumberland, Durham and Newcastle-on-Tyne, 1972)

Kirkaldy, J.F., *Fossils in Colour* (Blandford, 1970)

Lee, B.H. (Ed.), *Lead Mining in Swaledale*, from the manuscript of E.R. Fawcett (Faust, 1985)

Neruchev, S., *Uranium and Life in the History of the Earth* (USSR, 1984)

Park, R.G., *Foundations of Structural Geology* (Blackie & Son, 1983)

Postlethwaite, J., *Mines and Mining in the English Lake District* (Postlethwaite, 1913)

Price, R.J., *Highland Landforms* (Highlands and Islands Development Board, 1976)

Rastall, R.H., (revised by) *Lake & Rastall's Textbook of Geology* (Arnold, 1968)

Robson, D.A. (Ed.), *The Geology of N.E. England* (The Natural History Society of Northumbria, 1980)

Sherbon-Hills, E., *Elements of Structural Geology* (Methuen, 1963)

Sorrell, C. & Sandström, G., *The Rocks and Minerals of the World* (Collins, 1978)

Stamp, L.D., *Britain's Structure & Scenery* (Collins, 1984)

Weyman, D. & V. *Landscape Processes* (Allen & Unwin, 1977)

Whittow, J.B., *Geology and Scenery in Scotland* (Penguin, 1977)

De Zanche, V. & Nietto, P., *The World of Fossils* (Sampson Low, 1979)

Index